Work, Unemployment and the New Technology

Colin Gill

Work, Unemployment and the New Technology

Polity Press

First published 1985 by
Polity Press, Cambridge, in association with Basil Blackwell, Oxford.
Editorial Office: Polity Press, Dales Brewery, Gwydir Street, Cambridge, CB1 2LJ, UK.

Basil Blackwell Ltd
108 Cowley Road, Oxford, OX4 1JF, UK.

Basil Blackwell Inc.
432 Park Avenue South, Suite 1505, New York, NY 10016, USA.

British Library Cataloguing in Publication Data

Gill, Colin
Work, unemployment and the new technology.
1. Labor supply—Effect of technological innovations on
I. Title
331.12'5 HD6331

ISBN 0-7456-0022-0
ISBN 0-7456-0023-9 (pbk.)

Library of Congress Cataloging in Publication Data

Gill, Colin.
Work, unemployment, and the new technology.

1. Unemployment, Technological. 2. Work.
3. Microelectronics—Social aspects. I. Title.
HD6331.G55 1985 331.13'72 85-3643

ISBN 0-7456-0022-0
ISBN 0-7456-0023-9 (pbk.)

Typeset from data, by Graphiti (Hull) Ltd.
Printed in Great Britain by Billing & Sons Ltd., Worcester.

Contents

Foreword

In the latter half of the 1970s a new technology based on developments in microelectronics came to the notice of the general public for the first time. Sir Ieuan Maddock, former British government Chief Scientist, described it as 'the most remarkable technology ever to confront mankind'. The new technology is a 'heartland' technology; it can be applied to every sector of industry and commerce and few occupations will escape its impact. The 'silicon wonder' has not only stirred the public imagination, as witnessed by the growth in personal home computers, but is now seen as a major issue by industry, commerce, politicians, economists, academics and the news media. One major area of social concern is the effect that the new technology will have on work, skills, employment levels and work patterns. What is the future of work in the 'information age'?

This book sets out to take a critical look at how our working lives are likely to be affected by present and future technological innovation. It is a small book on a large and complex subject; I have inevitably had to be highly selective in my choice of subject matter and I can only hope that it represents a small contribution to what is, or ought to be, a matter of major public concern.

This book owes an enormous debt to a number of my colleagues. In particular, I would like to thank Anthony Giddens of King's College, who read my entire manuscript several times. He saved me from many mistakes, questioned many of my judgements, and tempered the dogmatism of my style. Elizabeth Garnsey, University Lecturer in Industrial Sociology in Cambridge University Engineering Department, and John Thompson of Jesus College also deserve special mention. I would also like to thank several of my engineering colleagues who brought to my attention a number of important literature sources and who patiently explained to me the complexities of the technology. None of them subscribes to the narrow views of job design about which I have been so critical throughout the book. Finally, I must acknowledge my own debt to the new technology. The word processor which I used to type the manuscript considerably shortened the time required to produce a final version of the book.

Colin Gill, University of Cambridge, January 1985.

Acknowledgements

The author and publisher would like to thank the International Labour Office for their permission to reprint material in this book.

Glossary of Abbreviations

AFL–CIO	American Federation of Labor–Congress of Industrial Organisations
ALC	Arbetslivscentrum (Swedish Centre of Working Life)
ADP	Automatic data processing
AUEW - TASS	Amalgamated Union of Engineering Workers (Technical, Administrative and Supervisory Section)
CAD	Computer automated design
CAE	Computer automated engineering
CAM	Computer automated manufacturing
CBI	Confederation of British Industry
CEDEFOP	European Centre for Vocational Training
CFDT	Confedération Democratique du Travail
CGT	Confedération Générale du Travail
CGT–FO	Confedération Générale du Travail–Force Ouvrière
CNC	Computer numerically controlled
DCF	Discounted cash flow
DNC	Direct numerically controlled
EDP	Electronic data processing
EEC	European Economic Community
EFT	Electronic funds transfer
EPOC	Equal Pay and Opportunities Campaign
ETUC	European Trade Union Confederation
ETUI	European Trade Union Institute
FF	Försäkringsanställdas Förbund (Swedish Union of Insurance Employees)
FIET	International Confederation of Commercial, Clerical, Professional and Technical Employees
FLM	Italian National Federation of Metalworkers

FMS	Flexible manufacturing systems
FRG	Federal Republic of Germany
ILO	International Labour Office
LO	Federations of Trade Unions (in various Scandinavian countries)
LSI	Large-scale integration
MITI	Japanese Ministry for Industry and Technology
NAF	Norwegian Employers' Confederation
NC	Numerical control
NEDC	National Economic Development Council
NTA	New Technology Agreement
OCR	Optical character recognition
OECD	Organisation for Economic Co-operation and Development
OTA	Office of Technology Assessment (USA)
PTK	Privattjänstemannakartellen (Swedish Federation of Salaried Employees in Industries and Services)
PTT	Postal, telecommunications and telephone services
QC	Quality circles
QWL	Quality of working life
SAF	Svenska Arbetsgivareföreningen (Swedish Employers' Confederation)
SCB	Swedish National Central Bureau of Statistics
SPD	Social Democratic Party (West Germany)
SPRU	Science Policy Research Unit
STU	Styrelsen för Teknisk Utveckling (Swedish National Board for Technical Development)
TCO	Tjänstemännens Centralorganisation (Swedish Central Organisation of Salaried Employees)
TUC	Trades Union Congress
UAW	Union of Auto Workers (USA)
VDU	Visual display unit
VLSI	Very large-scale integration
WRU	Work research unit

Introduction

The New Information Society

We constantly hear today about the 'new technology' and how important it is for our future. Most people had never heard of the phenomenon before the end of the 1970s, and their experience was limited to a small number of gadgets based on microelectronics, such as pocket calculators, digital watches and electronic games. They may have known of the existence of word processors as being sophisticated and very expensive forms of typewriter, but the average person had probably never seen such a device, let alone had access to one. Of course, many were aware that computers were increasingly being adopted by a variety of organizations because their bills, wage or salary slips, air tickets, cheques and bank statements contained an array of digital figures, but this hardly seemed to be of major significance to their way of life.

After all, the members of modern societies are used to new inventions, and there was no particular reason to suppose that this 'microelectronics' technology was any more significant than other technological inventions that had emerged before. Ever since the dawn of the Industrial Revolution there has been a deluge of technological developments which have resulted in new products and industries, larger markets, increased productivity, economic growth and a level of prosperity which earlier generations would never have believed possible. Those who had predicted that the replacement of human labour by machines would lead to widespread unemployment and human misery had been proved wrong. Why then should this technology be any different from what had gone before? Surely the development of the new technology should create even more wealth, new employment opportunities, and perhaps even the dawn of a new 'information' age of increased leisure where human beings would be liberated from the constant need to work in order to survive?

Microelectronics has already been extensively applied in many sectors of industry and commerce, although not in a manner visible to the public. Robots are already installed in a number of manufacturing plants, and many people have seen television programmes depicting these strange machines working continuously on production lines. These 'steel-collar workers' do not need to be motivated to work and never need to take tea-breaks, nor do they submit wage demands. As the costs of applying microelectronic circuitry continue to decline, robots and other forms of computer-controlled machinery will be introduced on a massive scale into manufacturing industry. Not only will microelectronics be increasingly employed in the development of more sophisticated factory machinery, but it will also be incorporated into the design of the products themselves.

Microelectronics will be used in a wide range of applications apart from manufacturing industry. The growth of clerical employment, which was a feature of the early part of the present century, has already gone into reverse, and offices are being increasingly automated. The banking system by its very nature lends itself to the application of information technology – with its requirement for vast amounts of information storage, processing and transfer – much of it in numerical form. As microelectronics is increasingly incorporated into products, manufacture, design and services, it will bring about radical transformations in society and in work organizations.

It is clear that modern societies are being quietly and pervasively altered by powerful technological changes. There will be no turning back. The seeds of fundamental change in society have already been sown. Our way of life is going to be radically altered, for better or for worse, just as it has been in the past by such technological advances as the steam engine, electricity, the internal combustion engine and air travel. No doubt other technological developments such as genetic engineering and lasers will also significantly alter our way of life in the future. The key question is whether the difference between previous technological developments and information technology itself is such that we can justifiably describe what is happening as a second 'industrial revolution'.

The Emergence of the Microprocessor

Thirty years ago, electronic devices depended on the use of a vast array of vacuum tubes and valves, which required a lot of space, were expensive to produce and inefficient in terms of the power needed for their operation.

The transistor subsequently replaced these bulky materials and thus enabled a whole new generation of products to appear on the market. This meant that electronic equipment such as computers, television and radio sets became much more compact. However, these devices still had to be wired together, and since each single piece of equipment might have thousands of components which had to be connected with one another, electronic manufacture still remained a complicated and costly operation, and even then the various products were relatively bulky. Subsequently, the development of the integrated circuit allowed the transistors to be 'wired' together in one minute 'chip' comprising a vast, complicated circuit containing not only transistors, but also other components – such as resistors and diodes. Today, circuits containing 100,000 components in a chip measuring 5 millimetres across are commonplace. Yet the process of miniaturization is still only in its early stages, and it has been calculated that by the end of the present decade chips containing as many as a million elements will be available.

Parallel with these technological developments, mass production methods have reduced the cost of the chips to a tiny proportion of their cost three decades ago. A simple illustration of how dramatic these reductions in cost have been is that during the past 15 years the power of computers has increased by almost 10,000 times, while the price of each unit of performance has decreased 100,000 times. The computer of 1963 required several tens of thousands of hand-made connectors, all capable of failure. Today, the equivalent device would only require ten elements in a large-scale integrated circuit. It has been predicted that the largest computers constructed so far, which include hundreds of thousands of logic gates and memory capacities of over 32 million 'bits',[1] will, by the end of the century, be able to be contained in a shoe box and will cost around £700. An analyst at Massachusetts Institute of Technology (MIT) estimated that if today's computers are compared to cars in terms of cost and technology, the production of a Rolls-Royce would cost US $2.50 and it would run for 600,000 miles on a gallon of petrol.

Information Technology and Microelectronics

Alongside the developments in microelectronics, other related innovations have taken place which do not necessarily involve the application of microelectronics in an end-product. These include a range of technologies which, taken together, can be called 'information technology'.

When we speak of *information technology* we are referring to technologies relevant to human communication processes and to the handling of the information conveyed in these processes. In the broadest sense of the term, information simply relates to any and all facts that are communicated, learned or stored. Information science is now understood to mean the science by which computers process and store information; information technology is concerned with how computer-based information is stored, processed and transmitted – be it by telephone lines, cable, satellite, teletext or other means.

A key characteristic of modern information technology is that advances in microelectronics have gone hand in hand with innovations in telecommunications. Thus the fusion of information *processing* (represented by the role of computer technology) and communication (increasingly dominated by telecommunications) has brought about a revolutionary change in the quality of information *flow*. This means that more communicable information is available than ever before, and it can be processed much more efficiently and flexibly. It can also be transmitted and acted upon more rapidly.

Given the importance of information flow to the social infrastructure, the fusion of computers and telecommunications can be expected to bring about radical changes in the organization and administration of everyday social life. For example, information technology is applied in such areas as printing, telephone networks, broadcasting, and satellite communications; it involves the use of lasers, optical fibres, voice and speech synthesizers, infra-red light, sensory devices and so on.

Just as the Industrial Revolution affected the very fabric of society itself by shaping every form of human communications, so too does information technology. It does so by providing a storage medium for information and knowledge, much as the written word affected communication and human knowledge in earlier times. Not only does information technology provide a storage medium which is much more effective than the written language, but it can also extend communication from bridging time to bridging space. Its effects on society will therefore be just as significant as was the spread of printing on European thought in the Middle Ages.

We are moving increasingly towards an information society based on a microelectronic technology which is both capital- and labour-saving in its applications. The new technology is thus a 'heartland' technology which can potentially be applied in any sector of the economy.

Is Information Technology Just Another Technology?

Many social observers argue that the first Industrial Revolution took several decades to gather momentum and therefore could hardly be regarded as revolutionary in the political sense of the word. Nevertheless, it did at the same time mark an abrupt discontinuity in the evolution of both the economy and society – and represented a kind of mutation arising out of the technical changes at a particular point in time. This enabled a whole range of applications to be made across numerous industries. In this sense the microprocessor or 'silicon chip' could be seen as a means of providing the trigger point for another major discontinuity in economic development.

There are a number of reasons why the microprocessor can be seen as a revolutionary invention. In the first place, it has a characteristic which differentiates it from all previous technologies: it possesses an 'intelligence' function. Hence just as the steam engine transformed the limited physical strength of human beings in the production of goods and services, so too can the new technology enable human beings to extend their mental capacity to an unprecedented degree. Second, the new technology is universal in that there is virtually no field where it cannot be applied. The main factor limiting the pace of application is not the difficulty of identifying areas where microelectronics can profitably be employed, but the ability to develop and exploit actual applications. Moreover, it is significant that the microprocessor has appeared at a time when economic growth in advanced industrial societies can no longer be taken for granted.

The Social Effects of Telecommunications and Computers

As already mentioned, alongside the development of microelectronic-based computers, there have been dramatic advances in telecommunications technology involving the use of satellites, laser technology, optical fibres, videotext and other inventions. The marriage between computers and telecommunications considerably enlarges the social impact of information technology. Just as the telephone, radio and television enormously enhanced communications in society, so too will the fusion of telecommunications and computers expand the scope for information interchange.

It is probable that within two or three decades, especially in the industrialized countries, personal computers linked to a television set at home will become the central means for providing access to a wide range of services. Through this equipment, people will be able to order consumer goods after they have ascertained their bank balance on the screen, and they will be able to scan through a wide range of electronic mail, including newspapers and periodicals to which the household subscribes. They will also have immediate access to a wide range of databanks of knowledge, and the possibilities for education and undertaking training in particular skills will be considerable. Leisure-related activities may well be pursued in the form of sophisticated computer-based games involving friends and relatives at a great distance, and for those with musical tastes there will be a large repertoire of items to choose from.

It is unlikely that people will orientate their lives completely around the computer, given that the human need for social contact with others is universal. Prospects of a 'cashless society' with nearly all human communication taking place through telecommunication networks, and where voting in elections and referenda is available from the home, will be technologically feasible, but are unlikely.

Extrapolating from the Past?

It is often claimed that the modern microelectronic-based computers represent no more than an 'upgrading' of the early computers which were introduced during the 1950s and 1960s. Supporters of this view argue that fears that the early computers would cause large job losses and increased formalized control over work processes were subsequently shown to be unfounded, and that there is no inherent reason why the present generation of computers should have any detrimental effects on employment.

When the early automatic data processing systems were first introduced, they were applied to areas such as banking, insurance, accounting, statistics, record-keeping, social security, driving licences, tax collection, to name but a few. Such applications necessitated central data storage in order to take advantage of the economies of scale brought about by automatic data processing. This meant that the large computers of the 1950s and 1960s had to be located in close proximity to each other so that the various administrative agencies could make full use of them by means of time-sharing. Consequently, the large organizations which first used the early computers had to centralize their decision-making in order to make full use of automatic data processing.

Since then, the dramatic technological developments in the field of microelectronics, telecommunications, satellites, optical fibres, lasers and sensory equipment linked to computers have transformed information storage, processing and transmission. The automatic data processing that takes place in organizations today is considerably more flexible than during the 1950s and 1960s.

As we saw earlier, information technology is not 'just another technology'–it is a *revolutionary* technology. It would be dangerous and misleading to forecast the effects on work and society of the present generation of microelectronic-based computers solely on a simple extrapolation from experience of earlier forms of computerization.

Who Benefits from the New Technology?

While the advent of the new 'information age' will undoubtedly bring enormous potential opportunities for raising the quality of life, there are real dangers that social inequalities could be extended. The opportunities which open up as a result of the linkage between telecommunications and computers will depend a great deal on who has access to the wide range of databanks which will be created, and how these databanks and their associated monitoring systems are applied to the workplace and in society generally.

If past experience is anything to go by, the benefits arising from new technological developments will tend to favour those who are at the forefront in the investment and application of scientific and technological innovation. Internationally, countries in the Third World will be at a severe disadvantage compared to their counterparts in Western Europe, the United States and Japan. Projects concerned with social need will take second place to military developments; the most educationally advantaged will benefit more than the least educated; the research efforts of large corporations will receive more government support than that of universities; and those with the greater resources will derive more from the technology than those who are economically weaker. Information systems are usually designed to meet the needs of a small group of privileged specialist users. If this scenario is played out, there is a danger that skill differences will be widened, and the less privileged will be further disadvantaged compared to their richer contemporaries. Moreover, access to the new facilities will inevitably depend as much on education and skills as on the ability to pay for the new services.

The linkage between telecommunications and computers will enormously expand the flow of communication traffic across national boundaries. Quite apart from the obvious function of increasing the amount of espionage carried out by the major powers, trans-world data traffic will increase in relation to commercial and financial transactions, in the area of multinational corporate management and in the general sharing of satellite-based data management across the continents. Any assessment of the likely prospects of the peaceful uses of trans-world data flows must necessarily be speculative, but the richer countries will stand to benefit more than poorer countries.

Increasing Centralization of Power in Work Organizations?

Many experts argue that the era of large work organizations is now coming to an end. As we will see in chapter 1, the incorporation of microelectronics into products, manufacturing and services, coupled with developments in telecommunications technology, is likely to lead to smaller and more flexible work organizations. It is rightly pointed out that the new technology permits a *choice* in terms of how it should be applied. However, while the advent of small computers will undoubtedly provide *opportunities* for decentralizing power downwards in work organizations, arguments based solely on technical feasibility and economic factors alone will not necessarily ensure that there will be more autonomy and power at lower levels of work organizations. The possibilities for centralized control can still co-exist alongside increased freedom of action for groups and individuals at the lower levels. Those who hold the power at the top will still have the ability to monitor and control the work of those lower down in the work organization, and will be able to exercise that power effectively while at the same time conceding more 'autonomy' to those lower down. This stems from the ability of the top decision-makers to control the work being done at every level in the organization under their supervision.

It is a mistake to assume that small personal computers will necessarily help to limit the exercise of centralized authority. There are distinct advantages for the powerful groups in work organizations to leave employees free to make limited decisions on the basis of their expertise and judgement, but so constraining them through rewards, careers, promotions, appraisals, organization cultures, loyalty to the goals of the organization and through the careful design of information access. Thus a greater degree of decentralized decision-making can provide a smokescreen for more central direction.

Organization of the Book

The terms 'new technology', 'microelectronics' and 'information technology' are all interrelated. Indeed, these terms are frequently used interchangeably in everyday speech. As I mentioned earlier in this Introduction, the importance of *information technology* arises from the *fusion* between developments in microelectronics and telecommunications. The term 'new technology' can also refer to technologies such as bio-technology, genetic engineering and nuclear physics; these technologies, important though they are, are not within the scope of this book. To avoid confusion, where the term 'microelectronics' occurs in the text it can be taken to mean the miniaturization of electronics circuits, and 'new technology' will refer to the vast range of devices which are based on the link between microelectronics and telecommunications. The terms 'information technology' and 'new technology', for the sake of clarification, will be used interchangeably throughout the text.

Despite the fact that we are still at the beginning of the 'information revolution', there has been no serious public critique of the way product design affects people at work. This illustrates the extent to which we are still prisoners of the Victorian idea of science and technology which held that what is good for the industrialists and the major decision-makers in society is necessarily good for everybody else. Given the dramatic transformations in society and in the world of work that are now taking place, it is clear that we need to examine the implications of this revolutionary technology in a much more critical way than has usually been done up to now.

This book attempts to offer a critical view of the new information technology, concentrating on its likely effects on work and employment. The new technology is being introduced slowly but surely into workplaces throughout the industrialized world, and has already developed a momentum of its own. What effect will it have on work? Do we have to face the prospect of high levels of unemployment as automation is introduced? Will the new technology lead to de-skilling and increased routinization of work tasks?

Which occupations are most at risk? Although we are still in the embryonic stages of the 'information revolution' and many breakthroughs in telecommunications and computer technology remain to be applied in a social setting, this book seeks to provide answers to these and other questions. Inevitably, much of the discussion is of a speculative nature, if only because of the uncertainties of predicting the political, economic and technological changes that will take place in the future. Whatever form these changes take, the new technology presents opportunities and dangers both to society and the way work is organized. It need not necessarily follow any predetermined direction and it permits a multitude of *choices*; choices about employment levels, job design, skills, work patterns and the relationship between work and leisure.

The first chapter looks at the radical changes that are likely in the way that enterprises carry out their activities, stemming from the application of microelectronics to products, manufacturing and services. Chapters 2 and 3 discuss the effects that new forms of automation based on microelectronics will have on skills, employment levels and work patterns in office work and in manufacturing. Chapter 4 assesses the possible unemployment threat posed by information technology and examines the likelihood of new job creation in various sectors of the economy. Chapter 5 assesses the trade union response to the new technology.

Some of the most interesting ideas for alternative forms of job design and new methods of ensuring that technology meets the needs of employers, unions, consumers and society generally can be found in the Scandinavian countries; these countries warrant special attention in chapter 6. The theme of the book is that the nature of the new information technology is such that not only will it lead to radical transformations in work organization, but it will also represent a fundamental challenge to traditional attitudes towards work in modern industrialized societies. Some of the ideas that have been suggested about the future of work – and how the new technology can be used to the benefit of everyone – are explored in the final chapter.

1

The Impact on The Enterprise

We have already noted that information technology is a 'heartland' technology in the sense that it will effect radical transformations throughout society and in the world of work in particular.

There is no doubt that this technology is destined to spread to a vast range of economic activities, even though we are not sure how fast this will happen. It will lead to the emergence of new products and to new forms of work organization. On the one hand, it will lead to changes which will save capital, labour, materials and energy; on the other hand it could bring about an even greater use of these factors of production. Nearly every enterprise will be affected in one way or another by the new technology, regardless of its activities. Its impact will spread across the entire structure of an enterprise from manufacturing to administration, and from planning to marketing. It is just a matter of time.

Changes in the Economic Climate

In the introduction we remarked on the decline in economic growth throughout the Western economies from the early 1970s onwards. Enterprises have had to operate in a climate of general economic instability, frequent and unpredictable cyclical fluctuations, reduced productivity, rising costs of production, stiffer competition and increased costs of energy following the oil price rises in 1973-4. All this led to increased pressures on profits and cash flows which limited the scope for investment. In such a changed economic climate there is a strong incentive to attempt to offset the effects of inflation and high energy costs by seeking to concentrate resources on labour-saving techniques and radical transformations in company organization. The adoption of new technology thus provides an attractive means of ensuring the survival of the enterprise. It can be incorporated into new or considerably modified *products* such as digital electronic controls on various types of machinery, word processors and new office technology, traffic system controls and so on. It will also have a considerable influence on the provision of various *services*; finally, microelectronics will be employed in the *manufacture* of traditional goods. Whichever of these three forms are considered, it becomes clear that there will be substantial changes in the way that enterprises operate.

Microelectronics in Products

As we saw earlier, the basic element of microelectronics is the so-called 'chip' – a silicon wafer measuring a square centimetre or less which contains a multitude of transistors, integrated circuits and other components. At the beginning of the 1970s the component density of each of these chips was around 10,000 and will soon reach a million. The advantages of microelectronics are obvious: more powerful, much smaller, more reliable, reprogrammability, much cheaper, and economical in terms of energy consumption. Microprocessors are now used in more and more consumer durables such as household appliances, entertainment products and personal products, such as calculators and watches. They are increasingly being applied in the field of industrial control technology by replacing electromechanical equipment and control systems (such as temperature control and traffic control). In factories they are increasingly employed in computer numerically controlled machine tools, robots and materials handling equipment. They are also being used in public enterprises, including social security, tax administration, aerospace and defence, education and health. They are also used in the manufacture of computers, data-transmission equipment and telecommunication systems.

The new products are nearly always superior to the old in that they satisfy a greater variety of needs at a much lower price. However, whilst the new products are much more reliable and much cheaper, the great advantage of microelectronics lies not so much in the products and the processes themselves but in the fact that they can so easily be *linked together* to form integrated control and information systems for example linking the design, manufacture and inventory functions together in a CAD/CAM system to produce products at a much lower cost (see chapter 3) and linking word processors with electronic mail and filing systems to evolve into a 'paperless' or 'electronic office' (see chapter 2). The vast range of microprocessor applications can be seen in Box 1.1 which was compiled by J. Rada in a study for the ILO Geneva.[1]

BOX 1.1

Some of the More Common Microelectronic Applications

Domestic appliances: washing machines, sewing machines, ovens, mixers.
Domestic regulators: central-heating control, lighting control.
Leisure: radios and television sets, hi-fi equipment, tape recorders, video tape recorders.
Personal: pocket calculators, watches, personal computers, teletext.
Cars: dashboard displays, fuel supply and ignition systems, braking systems, collision-avoidance devices, diagnostic systems.
Telecommunications: radio and television, electronic telephone, telephone exchanges, telex-switching systems, data transmission, electronic facsimiles, electronic mail transmission, remote terminals, teleprinters, paging systems, teleconferring, electronic news gathering, satellites.
Office: accounting, computers, typewriters and word processors, copiers, computer microfilms, facsimiles, electronic archives and retrieval systems, recording, telephone-answering machines, dictation.
Trade: on-line ordering systems, automated warehousing and stock control, computer-planned distribution networks, point-of-sale equipment and terminals, automatic supermarkets.
Banking: automatic cash tellers, automatic transfer of funds, credit card systems, cheque processing systems.
Printing: linotype electronic systems, colour correction and storage, machine control.
Computers: mini-computers, microcomputers, input-output equipment, 'intelligent' terminals, optical and laser character readers, printers and displays.
Industry (general applications): measuring and test equipment, devices for monitoring dimensions, temperature, weight and other factors; plant and personnel control, electronic clock-in; robots (programmed and self-programming devices) for welders, carriers, painters etc.; monitoring and control of industrial processes in the

nuclear and steel industries, for high-risk operations and for mass production; machine tools; smelting, welding and electroplating equipment, textile machines, materials handling, etc.
Military and aerospace: air traffic control, radar system data processing, navigation systems, military communications, guided missiles, night-viewing equipment, microwave blind-landing systems, infra-red surveillance.
Design and construction: computer-aided design, civil engineering and related design and equipment.
Education: computer-aided education, general education techniques, audio-visual aids.
Health: filing, general hospital management, body scanners and advanced diagnostic equipment, heart pacemakers, patient monitoring systems, kidney dialysis equipment, computer-produced speech, electronic aids and sight for the blind.
Public administration: centralized filing, police filing, traffic control, mail processing.
Other: meteorology, pollution control.

It has been estimated that by 1986 the number of electronic functions incorporated into a wide range of products each year can be expected to be over 100 times greater than it was in 1977.[2]

Clearly, as microelectronics becomes increasingly incorporated into the vast range of products as outlined in Rada's list above, a considerable number of enterprises which currently use mechanical technologies will convert to electronics during the next few years. Microelectronics also serves to modify the boundaries between different industrial sectors by encouraging the convergence of different industries and enterprises. For example, the data processing industry and the telecommunications industry are already closely intertwined in most of the Western industrialized countries. This convergence will also have the effect of producing large conglomerates by a process of merger and acquisition; the telematics industry of the future will be highly concentrated in the hands of governments and large multinational companies.

Manufacturing engineers have been quick to note the advantages that the new generation of technology brings to the production cycle in manufacturing enterprises. The main advantage is that there is a consequent reduction of work phases in the production cycle. At the risk of over-simplification, this cycle can be summarized as a succession of transformation phases in which parts and components are made from raw materials, and sub-systems and end-products from assembled parts. Microelectronics makes it possible to build a growing number of previously separate functions and ready-assembled parts and systems into the components. As component integration increases, the number of phases of the manufacturing cycle is reduced. This is effected by what the engineers describe as LSI and possibly VLSI. As component integration increases, the costs of parts and various functions decrease, and this results in a shift of manufacturing activities and of value added from end-product enterprises to microelectronic components companies.

The effect of all this is that there is a reduction in the manpower requirements both of end-product enterprises and of microelectronic component manufacturers. Moreover, there will be a drop in the level of production integration and often a rapid decrease in the manufacturing activities of end-product production enterprises as a result of the increasing separation between the production of electronic components and the production of systems and end-products.

These changes in the manufacturing process can be illustrated by noting how many common household appliances have incorporated microelectronic developments in recent years, such as televisions, washing machines, refrigerators and tape recorders. All of these products now need fewer and fewer components in assembly. For example, digital electronic watches only require five components nowadays compared to over 1000 assembly operations for mechanical watches. Bruno Lamborghini gives a number of other examples:[3]

> . . . the production of an electromechanical teleprinter requires more than 75 hours, whilst an electronic teleprinter needed only 17.7 hours in 1980 (one microprocessor can replace up to 936 mechanical parts). An electronic taximeter requires 3.7 hours compared to 11.7 working hours for an electromechanical one. Over 9 hours are needed to assemble a mechanical calculator, whereas a printing electronic calculator requires less than one hour . . . in the manufacture of sewing machines one microprocessor replaces 350 mechanical parts.

The other effect of incorporating microelectronics in products is the more obvious one of serving to reduce manpower requirements quite substantially.

One particular industrial sector which is most affected by the advent of microelectronics is the computer industry itself. Here, enterprises are being radically affected by the replacement of highly labour-intensive electromechanical or mechanical parts with increasingly integrated components produced by automation by other component manufacturers. The office machinery sector is being affected even more, and the two sectors are gradually converging together. The new information technology thus greatly increases competitiveness and enables newcomers to enter the market and thereby modify the structure of product markets radically. Increasingly, there is an incentive for manufacturers of equipment to buy components rather than to make them themselves.

Economists would also point out that this process serves to increase industrial concentration. This will come about because end-product manufacturers in the information technology sector will seek to secure control of strategic component manufacturers by acquisition. Similarly, there will be an increased industrial concentration of power among component manufacturers and a migration of market control into the hands of large USA companies – on whom many information technology end-product manufacturers are so dependent. As Bruno Lamborghini states: Microelectronics tends to transfer the specificity of individual products to the applications for which the products are intended.[4]

The inevitable outcome of present technological developments is that eventually it will be possible to produce vast quantities of microprocessors with powerful memories, line controls and peripheral controllers all into one standardized 'black box' at minimum cost. The application of these 'black boxes' will depend on the particular software that is loaded into the 'black box' memory. The 'black box' will then be connected to the equipment which is designed to perform the specific applications which it controls.

There is a strong possibility that these 'black boxes' will be produced by a few very large manufacturing companies who will be able to reach exceptional economies of scale. Even though the end-product manufacturers will be able to benefit from the low costs of these multi-purpose 'black boxes', they will nevertheless be forced to customize their products and the ancillary software to the needs of the user in order to stay in business. This will mean that company structures will be radically altered as a result of the decline of manufacturing and the consequent increase in the provision of application software and customer services. In such circumstances there will undoubtedly be radical transformations in the managerial structures and the organization of their activities, and it is possible that many companies will consequently go out of business.

Such a trend will mean that fewer and fewer manual workers will be required and there will be a rise in the number of highly specialized and skilled technical personnel in enterprises.

One of the factors which limits the ability of companies to exploit the opportunities of providing personalized enhanced software application developments for their customers is the cost of the operating software itself. Thus at present there is an incentive to offer standard products and hardware because of the growing internationalization of markets. However, microelectronics does offer the future possibility of more and more customization rather than standardization. We are not sure exactly how this standardization v. customization dilemma will be resolved, but its importance to the future configuration of enterprises cannot be overemphasized.

Although we are still in the embryonic phase of the so-called information revolution, a number of challenges present themselves to management of enterprises in the near future as a result of the incorporation of microelectronics in products:

a The technological developments are proceeding so rapidly that the life-cycle of products is becoming much shorter than previously. This means that a premium is placed on flexibility in production and marketing planning.
b The need to locate industrial sites according to the most favourable labour and product market criteria will become less and less important as time goes on. Proximity to component supply centres and technological know-how will be more important.
c There is likely to be a diversification of business organization towards complementary mechanical and electronic activities in order to absorb overmanning and improve profitability, by creating a network of small enterprises which are highly specialized in specific fields.
d There is likely to be less emphasis on production itself with a greater commitment to pre-production (Research and Development), and post-production areas of the business – such as marketing, technical assistance and applicative software. All this will have profound implications for management structure and functional responsibility in enterprises.
e We can expect to see a continuous appraisal of 'make or

buy' decisions in enterprises because enterprise management will be aware of the risk of finding that component parts are unavailable – thus jeopardizing the success of an entire range of its products.

f The increasing importance of information in management decision-making is an inevitable consequence of advances in technological development. This will be true in the cases of incorporating microelectronics in products, the provision of services and in product manufacture. (This is explored separately later in this chapter.)

Microelectronics and the Service Industries

The service industries are considered in some detail in chapter 4, and microelectronics will have a considerable impact in this area.

Banking and insurance have so far been at the forefront of information technology applications; radical changes in their activities have already taken place with many more on the horizon. The use of teller terminals, automatic cash dispensers and EFT systems within and among banks has developed at significantly higher rates than in other user sectors.

There has been considerable intensification of the automatic handling of policies in the insurance sector, but it is in banking where most of the dramatic developments are taking place.

In Sweden for example,[5] one of the largest commercial banks, SE Banken, is moving towards complete automation of all its banking activities. A series of sophisticated front office terminals were due to be fully operative by early 1985 which would be linked to the bank's extensive computer network and used to provide more extensive services – including investment advice, loan calculations, and tax and financial analysis. The bank will eventually be able to provide computer programs to be run on at least 20 different types of microcomputer for customers' financial problems with direct communication to the bank's main computer system. Using a communications network, this will allow branches in remote parts of Sweden to contact financial specialists based at the headquarters. This will dramatically speed up the flow of information and advice out to the branches.

The pressures of competition are forcing banks to increasingly sell their knowledge and expertise and provide more specialist services rather than the more traditional business carried out by banks.

Although the technological potential exists for the basis of the so-called 'cashless' (and 'chequeless') society, where the flow of money will turn into a flow of electrons around the circuits of computer networks, it is by no means certain that such a scenario will occur. Nevertheless, retailing services, communication services and administrative and office work organization will be considerably altered.

In retailing services, the main structural effect will be to favour the concentration of large chain and department stores as a result of such innovations as point-of-sale terminals, on-line accounting systems, and the total automation of inventory control, product pricing and restocking. There will, however, be scope for increased specialization of the smaller number of smaller stores that remain. As home terminals become more widely used, business organizations will have to reorganize their activities to fit in with 'tele-shopping' facilities.

The telecommunications sector and the services it provides is expected to expand considerably in the near future; the range of activities in telecommunications has been enlarged from telephone networks to data, text, graphic and image services. In organizational terms, the most significant change that has taken place has been in the tendency for large postal, telegraphic and telecommunication authorities (PTTs) to have their monopolistic positions challenged by alternative private services; for example the breaking-up of A.T.T.& T. in the United States, the privatization of British Telecom in the United Kingdom). Public bodies are already being forced by market and governmental pressures to make radical re-organizations in the running of their services by introducing new marketing and promotion techniques, better customer service, etc.

In the office and administrative sector we have already seen how this is expected to have a dramatic effect on the enterprise; separate attention is given to this in the next chapter. The prolific growth in paper-work which has been a feature of the growth in office and administrative functions during the twentieth century could eventually slow down significantly as it is gradually replaced by electronic data storage, processing and transmission. The 'paperless' office is by no means an unlikely possibility. Enterprises will need to be able to take advantage of the availability of very cheap microelectronic data technology by integrating the vast increase in information available into the planning, decision-making and control functions. In manpower terms, it is worth remembering that it has been estimated that only $2000 of capital investment per employee in office automation has so far been made – compared to $35,000 of capital investment for every factory worker. Thus the office and administrative sector is ripe for automation.

Microelectronics in Manufacture

Until the advent of sophisticated electronics and computers in the post-second world war period, the automation of manufacturing could be described as 'hard automation', where the manufacturing process was characterized by the use of stops, cams etc. so that human intervention was not required. This type of machinery had its origins in the nineteenth century, and whilst it continues to be important today, its use has largely been confined to mass and large batch production. However, automation of batch production is on the increase, and automation – even for very small batches or 'one-offs' – is becoming increasingly viable by means of electronic and computer-controlled machinery – where control of the production of a component lies not in the physical hardware of stops and cams, nor in the human machine operator, but rather in the programmable software of the machine's electronic control system. This, together with its effect on workers, is dealt with in much more detail in chapter 3.

What effect will this have on the management of the enterprise? CAD and CAM, whereby the design and manufacture of products can be integrated onto one computer control system, provides several problems for management.

First, there is a problem of what Arnold[6] has called the 'computerization' effect on management thinking. In a survey which was carried out for the Engineering Industry Training Board, most users of CAD found that they had to reorganize themselves to cope with a computer by making their procedures more explicit and to think (often for the first time in years) about their procedures and standards.

Second, because there is no user experience, management generally do not know which are the relevant criteria for use in system selection. Nor do they have much experience upon which to base cost-benefit analysis of CAD. In addition, management does not always realize that individual departments of firms cannot be simply treated as profit centres; nor can capital investments be justified on the basis of discounted cash flow (DCF) criteria. Both of these management techniques are inadequate in coping with new technologies such as CAD/CAM and flexible manufacturing systems. Such technologies offer systems gains, potentially improving the performance of the firm as a (whole) and not simply the one department which has to attempt cost-justification.

Third, skill factors are crucial in the adoption of new technology. Management skills are needed in combination with engineering skills to assess not merely returns on investment at profit centres, but systems costs and benefits. As Arnold notes:[7]

> Presently, career managers rarely have engineering skills – indeed their training usually has more to do with accountancy and law [not that these are in themselves unimportant.] Equally, engineers promoted into management are rarely trained in managerial and industrial relations skills.

Finally, as Francis has pointed out,[8] when companies introduce CAD/CAM and other forms of computer automated manufacturing into the enterprise, they may have to change their organizational structure to allow for a product-based form of organization rather than a functionally-based one. They may even need to adopt a 'matrix' form of organization.

In summary, microelectronics is expected to have profound effects on enterprises in terms of being incorporated into products, in its use in manufacturing and in its application to the services sector. It will require a major restructuring of enterprises, not just at the manufacturing stage, but at all levels. In certain industries, particularly in the growing area of electronics, there is an increasing need for high investment in research and development, which is of fundamental strategic importance. The pressures of competition, with the short life-cycle of products, will mean that companies must be able to keep one step ahead of their rivals in order to survive. More particularly, only if a company is able to lead innovation in a particular product can it achieve the level of earnings needed for reinvestment in R & D. On the other hand, in order to be a leader in innovation, sufficient resources need to have been invested in R & D. This 'Catch 22' situation increases competitiveness and the rate of innovation. This places a premium on the need for enterprises to forecast when the commercial production of the next product generation is likely to take place, and then design their products to coincide with the availability of the new components.

One of the major effects of the new technology is to permeate the whole of the enterprise. As we have seen, it is not simply the case that the new technology substitutes human labour with a form of automation, important though that is. It is the *information* part of information technology that is significant. Only if this information is appropriately used, can its flow enable enterprises to adjust to the challenges of increased competitiveness; failure to so do could have catastrophic consequences. Information technology also serves to highlight any inherent weaknesses in an enterprise's structure, as well as its managerial capabilities and its market position. This will be true whether the company is incorporating microelectronics in its products or whether it uses microelectronic-based equipment in its production, administration or service activities.

It goes without saying that the retraining aspects of the new technology for enterprises are immense. Paradoxically, there is a tendency for enterprises to place much greater emphasis on investment in technical change than on the necessary investment in training and education of the workforce to make full use of its capabilities and organizational consequences. In a paper by Sheila Rothwell and David Davidson of the Henley Centre for Employment Policy Studies,[9] a number of important findings emerged from a survey of 20 different enterprises.

a Organizational decisions to go ahead with the new technology investment were usually found to have originated in the enthusiasm of a particular senior company director – often from the marketing function. The training implications depended on whether or not this director was 'people-oriented'.
b The companies that took a broad approach to communicating and training were more likely to involve and consult those employees most likely to be affected. Information technology served to highlight the confusion and conflict that existed between the different departments, and there was a need to break down interdepartmental barriers – particularly in relation to 'management services'.
c There was a high dependence on sub-contracting training of employees to suppliers' training courses. In several instances, this caused problems for the enterprise when a particular subcontractor's key systems analyst left.
d Traditional supervisory skills of improvisation and short-circuiting the formal procedures were no longer possible. Instead, they were forced to adapt to the formalized and disciplined routines of computer planned schedules.

An interesting study which was very much in line with these findings was carried out by Patrick Dawson and Ian McLoughlin of the New Technology Research Group at the University of Southampton in the UK.[10] They found in a study of the effects of computerization on railway freight supervisors that

> computerization has enabled management to develop a strategy for redefining supervision in a manner which both increases headquarters control and enhances the role of local supervisors in the overall system of management control of freight operations. The latter has been achieved by the delegation of more responsibility for day-to-day operating decisions to local areas by the creation of a new supervisory post explicitly concerned with using the real time information generated by the computer to control operations. Similarly, the availability of this information has enabled a centralisation of control at regional and national headquarters by making local operating conditions and performance immediately visible to management.

Enterprises are also affected by information technology in terms of their marketing and commercial structures. Planning in marketing becomes much more closely associated with product planning and the design stage; as enterprises become less manufacturing-oriented and more service-oriented there is an increasing need to pay more attention to marketing and after-sales service.

In addition, enterprises incorporating microelectronic components in their products have had to develop complex 'make or buy' policies in order to cope with the speed of technological development and ensure maximum flexibility and adaptability to new products, not only for components but for intermediate and end-products as well. This means that purchasing policy is particularly important, since it makes it possible to reach an adjustment speed and cost levels which internal production often does not allow. In addition, as Lamborghini points out:[11]

> Company planning must constantly bear in mind the need for technical and manufacturing agreements, exchanges of know-how, patents and licences with other manufacturers, or mergers and acquisitions.

Small is Beautiful

A number of authors have noted the likelihood that organizations will become smaller as information technology is more widely adopted in work organizations. Until quite recently, many companies (particularly those in small batch engineering) were able to adopt a very simple organizational structure. It was quite sufficient to rely on the traditional skills of individual crafts workers and draughtspersons using their initiative under the control of some laid-down rules and procedures – backed up by the moderate use of hierarchical control through line management and supervisors. Staff personnel would advise line management at various points in the organization.

However, Francis has suggested that CAE will lead to new organizational structures within the enterprise. He notes that a popular fashion in Sweden, promoted by the Swedish Employers' Federation (Svenska Arbetsgivareforeningen) is to set up 'companies within companies'.[12] This can involve reorganising an area within the factory so that the design, planning and production facilities for one particular product are located under one roof – quite separate from other products.

> What is lost in economies of scale and from benefits of functional specialisation is claimed to be more than recouped through the increased administrative efficiency, faster response times and quicker throughput. In addition, benefits in working attitudes, morale and commitment are sometimes claimed.[13]

The realization of the 'small is beautiful' notion is linked with the implementation of computer automated manufacturing. It is likely that plants, possibly even large corporations, will shrink quite substantially in terms of numbers employed. Moreover, the availability of relatively cheap forms of automation by reprogrammable CAE devices[14] will enable enterprises to batch-produce items that were previously produced with hard-wired automation and to reduce the size of economic batches. Thus, in the not too distant future, smaller firms may be able to take on business that was previously carried out only by large companies.

Information technology will enable rapid co-ordination to take place throughout different management functions in the enterprise. Thus the need for the various stages of the production process to be situated under one roof, co-ordinated by a vertically integrated management hierarchy is no longer a necessity once the new technology is fully implemented. For example, Rumelt[15] drew attention to the way in which enterprises could be widely dispersed without any consequent loss in managerial co-ordination and control as a result of new technology. He suggested that when information technology is combined with CAD/CAM, whole segments of the production process could be spacially and organizationally separated. Once computers begin to specify the requirements for parts fabrication in a standardized language and once the automated facilities exist to turn these specifications into a part, the need to have all the stages of engineering within the same location, or even in the same enterprise, diminishes. Thus the problems of managing large production units would thereby be avoided. It is likely that CAE will enable vertically integrated operations to uncouple their production processes – leading to the emergence of many small single-product or single-function companies. As Francis points out[16] a logical extension of this is where workers, either employees or freelance, can operate their word processors, CAD systems etc. from home.

The Move Towards Homeworking?

Much speculation has taken place about the potential that information technology offers for the development of home working or some form of self-employment. A quick scan through job advertisements (especially in the data processing field) will provide several examples of jobs which can now be largely carried out away from the employer's premises. Homeworking has been traditionally important in certain industries such as clothing and textiles. The belief that such homeworking or 'outwork' is somehow an anachronistic carry-over from the past does not stand up to all the empirical evidence, as Rubery and Wilkinson[17] have shown.

The main attraction of homeworking for employers is financial: there is no need to provide factory or office facilities. In most cases lower wages or salaries can be paid because of the reduced ability of homeworkers to take collective action through trade unions. In any case, it is much easier to lay workers off when they are not based on the employer's premises. Information technology considerably expands the number of occupations that can be suitable candidates for homeworking – particularly clerical work.

In the manufacturing area, a major constraint which has prevented factory workers from carrying out their work at home up to now has been the need to operate large and/or complex machinery which needs factory backup facilities or the availability of rapid access to information. CAE removes many of these impediments.

Similarly, a large amount of office work can easily be carried out at home by means of computer terminals which are linked to a central computer via the telecommunications system. Work can therefore be carried out at home and be monitored and controlled in a central office. This obviates the necessity for exercising managerial control, which has always provided a major constraint for the expansion of homeworking. Many areas of female clerical work lend themselves readily to homeworking, and the consequent dangers of social isolation for women within the home are readily apparent.

Working from the home presents women with a number of advantages but it also has many drawbacks. Perhaps the most obvious benefit for women is that it increases the flexibility from combining work and motherhood. Despite the maternity provisions that have been enacted in many European countries to enable women to take maternity leave with pay from their employment for a specified period, many women find that they have to leave the labour force for a lengthy period to raise their children, and when they return they find themselves irretrievably behind in their career development. Whilst some may be able to work part-time, such part-time jobs are frequently characterized by low pay and inferior working conditions.

If present experience is anything to go by, the future for (women) homeworkers seems bleak indeed. The Low Pay Unit in Britain has compiled a mass of evidence of exploitation. For example, Andrea Waind[18] described a series of horror stories: one worker was paid 24 pence (per day) for making up handbags. One firm in London was paying homeworkers £5 for the 72 hours needed to knit an Aran jumper. In Leicester, the centre of hosiery outworking, the average pay was between 50 pence and 90 pence an hour (the Wages Council minimum being £1.50). Some 10 were earning *15 pence* an hour. New office technology could well lead to a growing number of female office workers being transferred to low paid homeworking.

Up to now, the expansion of service industries and office employment has gone hand in hand with the expansion of paid employment, particularly in the post-war years. Since services were less vulnerable to recession than other sectors of employment, they have afforded women a certain protection from unemployment in times of recession. With the advent of new technology, the advantages that women offer to employers – i.e. cheap and relatively docile labour – become less and less relevant. Moreover, the types of work that many women do – low-level, repetitive and monotonous – make them particularly susceptible to rationalization, both in manufacturing and services. The fight for jobs on the employer's premises in the future might well be a fight for (women's) jobs, and those who advocate that 'a woman's place is in the home' may well see such a scenario being realized.

Home working can certainly allow more flexible work scheduling around the needs of children, but there are a number of drawbacks that cannot be disregarded. Working in isolation results in difficulties in making contacts in the employing organization, the absence of feedback, anonymity and social isolation from other workers. Moreover, there is a strong likelihood that women will be exploited in terms of pay and benefits.

Higher levels of management are also likely to be affected by the trend towards homeworking. As was mentioned previously, the future development of information systems should eventually permit communication of a high quality over distances. 'Teleconferencing', which substitutes for the need to bring staff physically together for management meetings by providing video linkages, is still very costly and rather inconvenient at present. However, according to a survey of 255 among the largest 1000 UK companies, almost two-thirds believe that by 1988 they will be employing executives working from home.[19]

As John Child noted in a recent article in 'Omega'[20]

> Rank Xerox has initiated networking arrangements with some of its specialists whereby they now work from their own homes on individualized contracts, and often have Xerox 820 microcomputers linked to the company's head office. The saving to the company is reported to be substantial since it estimates that a manager's or specialist's employment cost approaches three times his or her salary once overheads, secretarial and office services and administrative back-up are taken into account. Under the networking system, payment is only for a contracted number of days and/or services rendered and not for the non-productive time contained in full-time employment. With such arrangements, the use of new technology increases the ability to record the networker's work output, which further adds to managerial control.

New Technology and Management Structure

Up to now, most of our discussion of the effect of new technology on the enterprise has largely ignored management itself. One of the features of economic development during the 1960s and 1970s in the Western industrialized nations was a general trend towards larger organizations as more and more mergers took place. The same period saw a growth in employment concentration, and the rise in the numbers of large, complex multi-plant companies which spread their activities across a wide range of industries. The problems of management organization in these organizations increased as companies grew in size and diversification, with the need to employ more specialist staff and to innovate in an increasingly competitive environment. Throughout this period, companies had to employ more and more managerial, technical and administrative staff with the consequent elongation and complexity of management structure. All this meant that companies were forced to give considerable attention to ways of improving co-ordination, integration and control so as to ensure that these giant companies functioned efficiently.

However, new technology is spawning novel ways of dealing with the problems of managerial complexity; the 'small is beautiful' philosophy and the move towards matrix forms of managerial organization are but two of the possibilities for the future. In fact, whilst the new technology creates greater *choice* for the structure of management organization and the exercise of its functions (this is discussed further in chapters 2 and 3), there is a general expectation that it will also permit the *contraction* of management hierarchies and a radical *simplification* of management structures.

One reason why many enterprises will be keen to take advantage of the new technology is that managerial overheads have increased substantially in recent years.[21] As Child points out,[22] information technology will enable managerial overheads to be reduced in three main respects.

First, office automation can lead to substantial reductions in administrative and clerical staff (see chapter 2); second, information technology will serve to enhance managerial control and integration and thus reduce the need to rely on supervisors for monitoring and contingency handling and to employ middle managers for processing information; third, the increasing use of personal microcomputers and terminals connected into networks will shed doubt on the continued existence of large central data processing departments.

Another reason why we can expect management hierarchies to shrink is that there is an increasing tendency for large organizations to *subcontract* many of their activities to other smaller organizations. This tendency is partly connected with new technology but it is also a reflection of a change in managerial philosophy generally. Subcontracting has always been practised by organizations and it has often been called different names. For example, Marks and Spencers in the UK have contracted out all their production; many manufacturing firms subcontract the manufacture of component parts, whilst retaining responsibility for design and assembly; and agencies have long been a method of subcontracting selling.

At the extreme, the core of an organization need contain no more than a design function, a costing and estimating function and a marketing function, as well as some co-ordinating management. Within the public services sector there have recently been moves to 'privatize' certain activities. The reason why such trends are emerging lies in the ability of small specialist groups in the service sector to run themselves more tightly. Unlike the manufacturing sector, economies of scale in services peter out very quickly because the bureaucratic burden becomes too great too soon. 'In production systems large is cheaper. In service systems small is efficient.'[23]

Despite the fact that new technology will permit organizations to be much smaller and obviate the need for elongated management structures, there will probably be a reluctance on the part of many large organizations to dispence with the advantages of economies of scale and size. Such organizations might well try to get the best of both worlds by organizing themselves on a 'federal' basis. 'Federal' organizations 'are composed of individual units that group themselves together for better co-ordination of their key resources (purchasing) and/or defence (marketing)'.[24] Nestlé of Switzerland and Philips of the Netherlands are examples of such federal organizations. Federalism is, of course, similar to the 'companies within companies' philosophy that is currently in vogue in Sweden.

Handy claims that ultimately the adoption of new technology will facilitate the emergence of what he calls the 'professional' organization. As he puts it:

> The repetitive bits of work will be progressively automated. The accessory bits to the core of the organization's work will be gradually contracted out. What will be left? The specialists, the experts and the co-ordinators with a few helpers and dial watchers. More and more the organizations of industry and business will come to resemble organizations of professionals.

Handy's predictions may be stated as follows:

a fixed-term contracts will become more common, as firms refuse to take on the responsibility of employing an expensive specialist for forty years or more;

b flexible time contracts will be introduced to make part-time work more feasible when an organization does not need the full-time services of certain specialists, many of whom will be pleased to be employed for part of their time;

c personal and portable pension schemes will become more common in order to facilitate the moves above;

d education and training will become an ever increasing priority in organizations, partly because training will be a way of harmonizing the interests of the individual being trained with the interests of the organization.

The scenario which has been outlined above may be an attractive one for senior management in so far as it enables managerial overheads to be drastically reduced. It may also be attractive for those relatively few key staff who will retain prospects of full-time employment and for those who are willing to accept the risks of working independently on short-term contracts. Unless we can find alternative ways of reorganizing work in the future to deal with the problems of large-scale job displacement (both managerial and non-managerial staff), this scenario will be an appalling prospect for those who are without any kind of employment at all as a result of technological innovation.

There are, however, grounds for believing that this scenario will not necessarily be fully realized. Large service organizations offer personal services of a kind which have to be adapted to individual needs (such as health, education, social services, hotels and travel agents). In these organizations new technology could be used to complement rather than to substitute for the work of such staff. Second, we may well be under-estimating the collective negotiating power of those managers whose jobs are threatened; many of them occupy key strategic positions within their organizations, and are unlikely to allow their interests to be sacrificed so easily.

Towards the Japanese Enterprise Model?

The phenomenal success of Japanese companies has been the envy of politicians, businessmen and the public in the West for a considerable time. Despite constraints on expansion in the form of uncertain energy supplies, labour shortages, fierce international competition and rising prices of most natural resources, the Japanese economy has demonstrated its resilience in the face of such obstacles. Whilst Japanese success is in part due to the positive role that the Japanese banking system has played in fostering industrial development by being willing to invest on a long-term basis without demanding, or expecting, short-term returns, together with the co-operative relationship between government and industry which has been fostered by MITI, many people have pointed to the co-operative relationships between management and employees at enterprise level as the key to Japanese success.

In particular, observers have identified the lifetime employment system which is a characteristic of large Japanese companies; the enterprise-based trade union organization; the remarkable commitment which Japanese employees exhibit towards their work; the system of seniority-based payment which is linked to job flexibility and the massive attention that is given to training. Space does not permit a comprehensive evaluation of the Japanese employment here, and the differences in employment patterns between Japan and the West are well described elsewhere.[25]

It is important to stress that the guarantee of lifetime employment only applies in the very large Japanese corporations, and even there such guaranteed employment generally ends at the age of 55. In addition, the large corporations are surrounded by a sea of small subcontractors, which use low-paid labour and do not offer life-time employment. These small companies allow the large enterprises to concentrate on design, marketing and final assembly and they absorb the fluctuations in demand in labour and product markets. Some 70 per cent of the output of these smaller businesses is taken up by the large corporations. Another point which is worth stressing is that Japanese culture exerts a strong influence in promoting co-operation between employees and their companies and also between the subcontractors and the large enterprises.

A number of commentators in Western Europe and the United States have speculated that employment patterns in the West are moving towards the Japanese model largely as a result of the effects of new technology and the recession. William Brown, for example, has claimed that the structure of British trade unionism in private manufacturing industry is particularly susceptible to a set of management strategies which tend to isolate workforces both from the labour movement outside the enterprise and from the external labour market itself.[26] Those companies that pursue strategies of training their employees in the particular package of skills appropriate to their own technologies, rather than simply buying in a time-served craftsman or starting a conventional apprentice are effectively limiting trade union control and enhancing the employee's dependence on the employer. As Brown states:

> The more an employer can occupationally isolate his employees, the more it becomes worthwhile to offer them inducements to stay with the firm, accept periodic retraining, and work with flexible job descriptions.

A cursory analysis of the Japanese employment system shows why the employees there are so motivated and adaptable. Japanese employees in large enterprises are almost completely isolated from the outside labour market, and interfirm mobility is almost unknown. Second, company bonus schemes which typically add on around a quarter to a half of the employee's basic salary depending on company performance, are best seen as devices whereby the employer can effectively (share risks) with his employees. Third, the system of individual assessment is such that there can be a large diversity in the range of salary levels which are paid to a group of newly recruited employees who all started at the same time. Finally, not everyone in the large corporations enjoys lifetime employment; such practices are almost exclusively applied to male employees and there are a number of employees in the same firm who are employed on contracts of under a month's duration.

There seems to be a consensus that the twin effects of the recession and new technology are considerably enhancing managerial control at the level of the enterprise, both in Britain and elsewhere in the West. As the *Financial Times* remarked:[27]

> Flexibility is the name of the game in the 1980s labour market – in wages, skills and training. It is fashionable among British and other European policy-makers to admire the relative flexibility of employment in the U.S., some of which is ascribed to that country's large, relatively low-wage, labour-intensive service sector, and its entrepreneurial drive.

Many European governments currently have multi-headed programmes to encourage the emergence of the service sector, and to turn the unemployed and others into entrepreneurs.

Interestingly, whilst the employment patterns in the West may be moving towards a form where they increasingly resemble those in Japan, there is evidence that the Japanese system is itself threatened by microelectronic technology. One author[28] suggests that despite the success of the Japanese style of management in facilitating an increasing degree of automation in Japanese enterprises, this style of management actually contains the seeds of its own destruction.

Nuki examined the effects of microelectronic technology on the Japanese seniority system, the lifetime employment system, enterprise unions and the structure of the Japanese firm. He points out that the seniority system is under threat because of the twin effects of lower growth and microelectronic technology. Lower growth means that the average age of the staff in Japanese firms rises considerably and the firm's costs rise with it. The microelectronics era places a greater emphasis on technological skills: these skills are more likely to be found among young Japanese staff, whereas the old system presumed a close link between seniority, skill and merit. Seniority is becoming less and less important as a basis for salary remuneration in Japanese enterprises.

The traditional system of permanent employment in Japanese enterprises meant that staff were taken on young and trained in the light of that firm's needs and practices. Horizontal mobility between firms was almost unheard of, and if such movement did take place the employee concerned would receive a much lower salary. With the increasing importance of microelectronics, smaller and medium-sized firms especially are being forced to attract the technologically based skills they urgently require by inducing employees to leave the larger firms with the prospect of high salaries. In order to remain competitive, firms are obliged to call in skills from outside and are no longer in a position to guarantee their old employees a permanent place. Some firms are seeking to make technological innovation and permanent employment compatible by experimenting with lowering the retirement age from around 55 to between 40 and 50 whilst retaining the same benefits that were previously applicable.

Advances in microelectronics have also forced firms to seek the services of experts outside their staff to carry out specific tasks. These experts are outside the enterprise union's control. Moreover, the work of some existing staff is becoming redundant as new technology is increasingly being applied.

This means that Japanese enterprises have to avoid dismissing their staff by transferring them to other firms that can make use of them. These staff still belong to their old enterprise union even though they no longer work there. Weak though these enterprise unions are by Western standards, new technology threatens to make them even weaker.

Perhaps the most important changes of all that are taking place in Japanese enterprises as a result of technological developments relate to the structure of the enterprise itself. Japanese firms have traditionally lacked any form of class consciousness since all staff are considered to form a single class – from the managing director downwards. Computerization carries a high risk of information leakage to outsiders, and in order to protect their data, sophisticated coding procedures are necessary to which only the very senior managers can have access. Information access therefore serves as a new form of status differentiation within the firm.

Another effect of computerization is that it cuts across one of the most important assets of Japanese culture in the enterprise, namely the way in which decisions are arrived at by an extensive form of human contact and participation, of which quality circles are perhaps the best known example. Computerization requires rigid and formalized procedures and thus the community spirit which lies at the heart of the Japanese system of work is increasingly under threat. As Nuki concludes in his article:

> The Japanese group spirit that was the springboard for the tremendous expansion that has taken place up to now is clearly an obstacle to the development of the creative spirit since it crushes any display of individuality in the bud.

Nuki also believes that the human dimension in Japanese enterprises will find its place only in the research and development field and will be replaced by machines in production and manufacture.

Conclusions

Information technology is destined to effect radical changes in the way that enterprises operate, not only in the Western industrialized countries but also in Japan. Hardly any area of the activities of enterprises will be unaffected. *Information* will become the predominant factor in production, marketing, design, and the whole range of managerial decision-making activities. Not only will virtually every occupation in the enterprise be affected, but the structure of management itself will be profoundly altered. It will be some time before the full impact of technological change will be felt. As we will see in chapter 4, much will depend on the *rate of diffusion* of the new wave of technological innovation.

In the next two chapters we will give more detailed attention to the impact of the new technology on particular types of occupations, concentrating firstly on the automated office in the next chapter, and then on examining the impact of new technology in manufacturing.

2

The Automated Office

In the last chapter we took a fairly speculative view about the impact of the new technology at the level of the enterprise. In order to make some sense of our arguments within a wider context, we have inevitably had to mention in passing some of the technological changes that are already taking place or are in prospect both within the office and on the factory floor. The next two chapters will deal in more detail with these two major areas of technological application, and we turn first to the office.

The Growth of Office Technology

We have already noted that the capital investment per employee in factory jobs has been considerably higher than in office work throughout this century. We also mentioned that as a result of technological developments arising from the marriage of automated data processing and telecommunications, the office sector is now ripe for automation, with the promise of dramatic increases in productivity.

Although the first commercial typewriter was introduced into the office as early as 1873, most of the important developments in office technology have only taken place in the last 25 years. Before 1873 the office was predominantly a male preserve and its function was largely concerned with the maintenance of accounts and the associated correspondence relating to accounts. Following the introduction of the typewriter, office functions expanded considerably and women were employed in the office for the first time. The typewriter enabled correspondence to be carried out over the whole range of business matters although there were few major improvements in typewriter design until the introduction of the first electric typewriter in a commercially viable form in 1935.

Another major change took place when the 'golfball' typewriter appeared in 1961, and this improved version of the typewriter provided wider opportunities to develop the concept of a memory function. Along with developments in the technology of electromagnetic tape, this enabled data to be stored so that it could be amended as necessary, thus enabling text to be corrected or reorganized.

However, such developments in applying a memory utility to typewriters were not really economically viable because the straight reproduction of text was much cheaper by using standard duplicating methods, such as carbon paper and methylated spirit. All this changed when the first electrostatic photocopier was introduced. The modern word processor can be traced back to parallel developments in computerization from the 1950s to the present day.

The telephone was preceded by the use of telegraphic systems to provide a means of communicating business transactions over a long distance, and has remained substantially unchanged ever since its introduction in Victorian times. The telephone network has also been used in the operation of the telex system, and in the last two decades the growth of national and international communications has been dramatic.

All these technological elements can be described in combination as office information technology, in the sense that they are concerned with the generation, processing and dissemination of information – which lies at the heart of all office-based white-collar work. As already mentioned, microelectronics is also bringing about a merger between the computer and communications industries – which is one of the reasons why the new technology is described as information technology. However, it is true to say that most communications still takes place by telephone, telex and the postal service, although with the development of digital telephone systems and the increased use of satellites for communications, computers become the means of communication, transmitting information and data many times more quickly than at present. In the near future, communications will be speeded up even more by the use of transmitting 'blocks' of information or data in bulk from one station to another and then distributing it through networks to the recipient – rather like the distribution of postal mail today. Hence, microelectronics has not only increased the capacity to store and manipulate data but also to *communicate it*. As one writer recently described this development:[1]

Indeed, creating, using, storing and transmitting information have

> become parts of a single network: information gathered at any point in that network can be processed locally, transmitted to a larger system, stored in a data base and then retrieved, reused, retransmitted in the amount of time it took formerly to type a letter.

Perhaps the most widely known type of modern office technology is the word processor, which enables text input to be captured into an electromagnetic memory bank and reproduced as many times as are necessary to allow for corrections, additions and alterations. The word processor operates just like an ordinary typewriter except that the machine automatically deals with line length and margins so that typing speed can be increased substantially. Nearly all word processors display the text on to a VDU screen so that it can be corrected before it is put into memory. Facilities on many devices now even provide for the text to be automatically checked for typing errors and spelling mistakes by checking the text against a basic 'dictionary' which is built in to the memory of the word processor. Cheap word processing chips can now be used in small personal home computers.

Word processors have several advantages over conventional typing. Quite apart from the obvious advantage of increasing the speed of document production by between 30 and 50 per cent, manual activities such as erasure, margin setting, and 'cut and paste' are eliminated. The editing stage is similarly considerably shortened and once the word processor is linked with communications equipment, documents can be moved much more quickly and reliably. If the word processor is interfaced with a photocomposition system, it can perform typesetting commands and thus eliminate the need for text and input commands in the printing room.

Facsimile equipment transmits source documents between remote sites using special receivers linked by a telephone line and it enables documents to be reproduced speedily and accurately over a long distance. However, such equipment is still costly, but it is likely to be replaced by communicating word processors in large enterprises before too long.

Another product of microelectronics that is now becoming available on the office technology market is optical character recognition equipment or OCR. This eliminates the need to key in text, which is one of the most time-consuming tasks in an automated system. The data are read directly from a source document by using a special font or typewriter golf ball which scans the document and converts letters and numbers to bits – thus making it readable to the computer or word processor.

If we add to this list of office technology equipment the facilities offered by computer output microfilm which considerably improves the storage and retrieval of files, the scope for automation in the modern office is substantial. Once all this new technology is linked with telecommunications equipment using optical-fibre cables and satellites, we are well on our way towards the 'electronic office'.[2]

Despite the array of new technology possibilities that are available as outlined above, the electronic office is still many steps away and many problems remain before it is realized. None the less, there is definitely a trend towards the electronic integration of office activities. Malcolm Peltu[3] cites the 'Dataland' experiments that are currently taking place at the Massachusetts Institute of Technology (MIT) which are funded by the US Department of Defense. Here, communication with computers is done with special 'joysticks' and pads in the arm of a chair and with voice commands and electronic pens.

Office Technology: How Rapidly will it be Introduced?

The factors encouraging the rapid introduction of office technology will depend on the low level of capitalization in offices; the cheapness of the technology; the low productivity of office workers compared with production workers; the soaring costs of office overheads; and the increasing competitiveness in the economy which encourages organizations to rationalize their activities wherever possible. There are, of course, factors which limit the diffusion rate of office technology. Several writers have stressed the *conservatism of management* itself as an inhibiting factor, whilst others[4] have pointed to the reluctance of managers to relinquish the status of having their own personal secretary; the wide distribution of secretaries in separate offices within an organization will serve to add to this reluctance in many cases. Managers might also be reluctant to accept new office technology until an executive can enter and extract information from the system without having to use a keyboard. Moreover, the 'teething' problems associated with new computer-based systems can be extremely frustrating.

There are also technical and economic reasons which might well limit the diffusion rate of office technology in Britain.[5] For example, electronic information systems present a number of problems connected with the legality and security of business information and correspondence which currently require some documents to be on paper; the need to keep parallel, paper-based systems going is a deterrent to potential users. Another reason cited was the lack of a telematic infrastructure (although as we noted in the Introduction to this book such an infrastructure will be set up before too long).

On the supply side, many commentators cite the lack of skilled workers and limitations on saving and investment which creates capital shortages.[6] On the demand side, there is not only a lack of awareness about the potential applications of the new office technology among many managers but there are also financial considerations. In Britain, as Werneke noted, the Advisory Council for Applied Research and Development found that the reasons the United Kingdom was lagging behind in the introduction of new technology were: a general lack of awareness; the lack of confidence to invest; shortages of appropriate staff; and the lack of economic stimulus.

Similar constraints have been noted in the United States[7] and one can also cite the deluge of literature that is currently appearing both in the United States and elsewhere on the subject of introducing new technology as having the opposite effect from what was intended, in that many managers feel reluctant to go ahead with introducing the new technology because they are so confused about the various products that are available and the incompatibility between one system and another. Premature investment in equipment that can become quickly obsolescent can be very costly.[8]

Predictions about the diffusion rate are notoriously difficult, but the indications are that there will be steady, if undramatic change as office technology is introduced. So far it appears that office modernization is proceeding in a rather piecemeal fashion and this has major implications for skills, job content, and labour displacement among office workers.

The Effect of New Office Technology on Skills

The entry of women into the office coincided with a massive expansion in that sector which is broadly described as clerical/administrative, but which, historically, can be seen as the beginning of 'management'– a process which involved the separation of conception and execution, or the mental/manual division of labour.[9]

BOX 2.1

Franco de Benetti, managing director of Olivetti, the transnational office equipment company, discussed the potential of introducing Taylorism into the office environment at a 1979 conference on office technology in London sponsored by the *Financial Times*:

> The Taylorisation of the first factories . . . enabled the labour force to be controlled and was the necessary pre-requisite to the subsequent mechanization and automation of the productive process. In this way Taylorised industries were able to win competition over the putting-out system. . . . Information technology is basically a technology of co-ordination and control of the labour force, the white-collar workers, which Taylorian organisation does not cover.

After this explicit admission about the design and purpose of micro-electronically-based office machines, he continued:

> However EDP [electronic data processing] seems to be one of the most important tools with which company management institutes policies directly concerning the work process and conditioned by complex economic and social factors. In this sense EDP is in fact an organisational technology and, like the organisation of labour, has a dual function as a productive force and a control tool for capital. . . . To sum up: the easy adaptability of work to the machine, the diffusion of equipment thanks to technological developments, the measurability of the improvement obtained and finally the increased power which the manager acquires are the cause of the exceptional diffusion of the mechanisation of the office.

Source: D. Albury and J. Schwartz, *Partial Progress: the politics of science and technology*, Pluto Press, London, 1982, p. 151.

'Taylorism', or 'Scientific Management', which was developed by Frederick Taylor in the latter part of the nineteenth century (see chapter

3), provided a major stimulus to the sophistication of management techniques.[10] This system involved the fragmentation of the tasks inherent in a job. Each subdivided task was then timed and paced, permitting management to guarantee the time the job would take and hence calculate labour costs. Tayloristic forms of management control have been largely applied in manufacturing industry rather than office work up to now, and even though there has been a great deal of office mechanization taking place, fully developed forms of Taylorism in the office environment are almost unknown. Nevertheless, Braverman did claim that there had been a universal tendency towards deskilling throughout the twentieth century, and office work was no exception. Is there an inherent feature of new office technology which inevitably leads to deskilling?

It is true to say that there are some deterministic features of new office technology – changes which will almost always occur when the technology is introduced. However, this should not be taken to mean that there will be a (uniform) set of changes in response to technological innovation in the office. Rather, different consequences for work methods, people and organizations will result from different techniques of designing and implementing the new systems. The main effects of office technology on office work relate to skills, formalization of office routines and polarization of jobs.

There appears to be a consensus that clerks and secretaries will find that the new office technology will require them to spend more time on monitoring or supervising processes which will be carried out by computer-based devices. A premium will be placed on an ability to exploit and evaluate the capabilities of computer systems rather than on traditional skills such as calculating, preparing statistics, typing and laying out a letter or spelling words. Significantly, the EEC Centre for Vocational Training[11] found that the skills needed to work with microelectronic-based production equipment meant that the training time for new operators was (lower) than with the old system, and it was suggested that the same could happen with computer-based office work.

One reason why one might expect some reduction in skill as a result of new office technology is that computer-based systems require the use of formalized.logic and predetermined programming which can reduce the choice available to individual employees. This need not necessarily be so providing that the new technology is designed to be flexible and adaptable to individual requirements. However, such user-oriented design is not always possible because of the inherent predefined codes, programmed procedures and other structural rules that are necessary in order to achieve the full benefits in efficiency that computerized systems bring to the office. Moreover, many authors have pointed to the possibility of 'Tayloristic' philosophies of job design – whereby a large number of very routine, low-skilled jobs were created for workers on assembly lines with the more skilled and challenging tasks moved to specialized.departments – being carried over to the office environment. Michael Cooley, a former chairperson of the Lucas Aerospace Shop Stewards' Combine, has described the introduction of the computer into the office as 'the Trojan horse of Taylorism' because it is used to spirit into offices the techniques of the assembly line.

BOX 2.2

Deskilling can occur in a variety of ways. In the midwest headquarters of a multinational corporation in the United States, secretarial jobs were broken down into component parts when word-processing equipment was brought into the department. As a result, one woman does electronic filing all day, another extracts data all day, one answers phones all day, another handles correspondence all day, and so on. The company requires that each woman complete a 'tour of duty' of several months in each subtask in order to be considered for promotion.

A recent study of Wall Street legal secretaries' jobs illustrates another way that skills are stripped away when technology is brought in and combined with new levels of job specialization and centralized administrative controls. Legal secretaries' jobs have been among the most prestigious and highly respected of clerical occupations. But Mary Murphree of Columbia University and City University of New York (CUNY) found that:

> While early forms of office computerization served to upgrade and assist secretarial worklives . . . current innovations are striking at the heart of the traditional legal secretarial craft and creating a number of problems. The most challenging and responsible tasks traditionally in the secretarial domain are gradually being transferred away from the secretaries to *cadres* of professional and para-professional workers such as para-legal assistant, librarians, accountants, personnel spec-

ialists and word-processing proof readers, thereby reducing the secretarial function to one of merely telephone gatekeeper.

Murphree found that in the firms where new office technology had recently been introduced, job responsibilities were split up and down. Desirable work and new skills were assigned to other workers, as well as new means of performing tedious tasks. Specialized departments were created, with the work-flow centrally monitored and co-ordinated. Many secretaries resented the sophisticated dictating equipment and centralized transcribing pools increasingly present in large firms. Together, these technologies made many prized and hard-earned skills, such as speed stenography, irrelevant. Secretaries were left in a tenuous position – still holding high status jobs but no longer using many of their skills and fearful of their future as the law firms continued their effort to phase out traditional secretaries, using the administrative cluster as a 'compromise' or 'interim' solution.

Source: Mary Murphree, *Rationalization and Satisfaction in Clerical Work: a case study of Wall Street legal secretaries,* cited in *Labor and Technology*, Department of Labor Studies, Pennsylvania State University, 1982.

In addition, the loss of traditional secretarial skills can lead to a greater degree of job polarization in offices. As Bjørn-Anderson puts it:

> Office work generally has a woolly, changing and informal character which, to a large extent, relies on the adaptive skill of the staff to cope with new situations. If office routines require limited skills and are tightly specified, the system could become too rigid and staff could lose the ability to handle the non-standard or non-programmed tasks which constantly arise in offices. Highly structured jobs with a low skill content are also those which are most easily eliminated through automation. The polarisation of office work could therefore lead to the complete replacement of many secretarial jobs by programmed computers.

One of the changes in work organization that frequently accompanies the introduction of new office technology is the splitting up of typing work and administrative work. In general, administrative tasks are ranked as being more highly skilled than straight copy typing, although it is fair to point out that copy typing does require a lot of skill in controlling the layout and presentation of the work. If a separate word processing department is established in an organization in order to do the bulk of the typing work, the new machinery can be operated much more intensively and thus maximum use can be made of the capital equipment. The result is that the word processor operator's job is deskilled since it is not only stripped of more varied administrative roles, but it no longer requires much skill in the accuracy, layout, execution and correction of the work.

Case studies have shown that good and experienced copy typists have felt a deterioration in the skill levels of their jobs with the introduction of word processors, whereas younger typists have felt an increase in their skills – apparently because they have the opportunity to operate a complex piece of machinery.[12] It also seems that many word processor operators become quickly disillusioned and dislike the increased work specialization once the novelty of using the new equipment wears off.

Despite the fact that the introduction of new office technology does create jobs that require new skills, the problem seems to be that the skills made redundant by the new technology are, generally speaking, not appropriate for the the new opportunities that are emerging. Clerical jobs such as typing, stenography, filing and book-keeping that the new technology will affect will require skills that are often low-level, task-specific and often restricted to the particular employer or enterprise. Thus there is a real danger that many clerical workers will be isolated from the external labour market because their skills will be company-specific. In contrast, the jobs associated with the *management* of the new technology require higher-level skills of a more analytical nature in order to design and analyse information in such a way to fit in with the objectives for which the information is intended.

The Effect of New Office Technology on Job Content

There are two conflicting views about the effect of new office technology on job content and work organization in clerical employment. Some observers argue that the use of new technology will enhance jobs and upgrade the skills of those involved in office work. According to this view, new technology can reduce the tedium of routine tasks and the monotony of jobs by allowing more variety and interesting activities to be undertaken. It is frequently claimed that secretaries and typists welcome the opportunity to use word processors or 'display-writers' and there is as yet little evidence to suggest that there is widespread antagonism on the part of clerical workers generally to new office technology.

On the other hand, the pessimistic view is well represented in the literature on new technology, with its adherents claiming that office technology will cause jobs to be increasingly routinized and progressively deskilled and fragmented. Whereas speed and accuracy were once the hallmark of a good typist, word processors enable a well-presented document to be completed without a high level of typing ability. Supporters of this view argue that new office technology can lead to a loss of job autonomy and independence, with a tendency for work to become increasingly mechanized and subordinated to the technology – rather like work on a factory assembly line.

As we will see in the next chapter, one of the early consequences of Taylorist philosophies being introduced onto mass production assembly lines was the guarantee that at the very least such forms of monotonous and repetitive work would be rewarded by higher pay. Interestingly, data from the United States[13] found that 'women clericals were doing more work for less pay, faster and for more people at once'. A 1979 management survey showed that full-time VDU operators earned only $7 more per week on average than conventional typists. Yet consultants estimated productivity gains ranging from 50 to 500 per cent when VDUs were used, depending on the type of document production involved. Moreover, in some 'sunbelt' cities, where office automation was being installed at the outset, VDU typists actually earned *less* than conventional typists.

The outcome of new office technology depends on the way that work is reorganized when new technology is introduced. Certainly there is ample scope for deskilling and degradation of office work, and that is why the trade unions, particularly in Europe, have stressed the need for participation and consultation whenever new office equipment is introduced.

However, there are some instances where new office technology has led to a greater degree of job enrichment. Werneke for example,[14] cites several examples from the American literature. At Citibank, the introduction of minicomputers resulted in the creation of a number of 'super clerk' positions. One woman reported that before the computers were introduced she performed only one small part of security transfers. Afterwards she 'did 20 different things' and was thus able to deal with all aspects of a particular client's needs.[15] At other companies there are examples of new office technology improving both the productivity and creativity of office workers.

Yet such increases in job enrichment stemming from the broadening of job tasks following the introduction of new technology do not necessarily guarantee that the resulting work represents an improvement in the employee's position. For example, in another American study which was carried out by Glenn and Feldberg[16] in the customer service division of a large public utility, it was found that the use of new technology resulted in closer monitoring of the work. Before the new system was introduced, each clerk provided customers with information on a specific area, such as sales, bills, or service/repair. Each job was specialized.and a customer call might have to be referred to several clerks before a complete answer was received. The new system involved centralized information, giving one group of clerks access to all aspects of service by using a computer terminal so that although the jobs were broader in scope, each job became more closely monitored because the computer system provided a tally of all the information processed. Moreover, each part of the job had to be done according to standardized procedures and within a particular time. An interesting feature of this study was the fact that the reorganization of job content meant that the job changed from being classed as one suitable for men to one which was more appropriate for women. This 'feminization' of the job resulted because 'specialist' technical knowledge was no longer required and therefore women could take over the new jobs after the system was introduced.

The same authors also found evidence of feminization of work as a result of new office technology in a study of the occupational structure of the insurance industry in the United States. There the fastest growing occupations were the professional and technical jobs – most of which were occupied by men. Following the introduction of new office technology there was an increasing polarization of skills taking place, with fewer and fewer traditional jobs such as record-keeping, policy payments and record assembly remaining – most of which were occupied by women. Other positions involving accounting and correspondence remained stable and there was a substantial increase in the number of secretaries. At the same time there was a pronounced change in the sex composition of the labour force with the more skilled computer-related occupations being filled by men.

In the United Kingdom one of the most important studies so far on the effect of office technology on jobs was that carried out by Emma Bird[17] for the Equal Opportunities Commission in 1980. In a study of nine companies she found that the introduction of word processors had brought few changes to the job of the personal secretary because they had generally been introduced into existing typing pools where there was a clear distinction in existence between copy typists and personal secretaries. In eight of the companies she studied full-time word processor operators were employed and these operators were not downgraded secretaries but copy typists. However, Bird found that the introduction of word processors had reduced the amount of time spent typing by personal secretaries (in one company the proportion of time secretaries had spent typing had dropped from 18 to 12 per cent) and thus many of those in all the companies studied found that they were noticeably under-employed compared to before. As a result, three of the case study companies planned to reclassify secretaries as administrative assistants, responsible to more than one manager. The women concerned did not feel that this represented an enrichment of their jobs because they had lost the personal status of being assigned to one particular manager.

On the basis of the small amount of empirical work that has been carried out so far, coupled with some degree of intuition, it would seem that it would be unwise for managers who are contemplating the introduction of new office technology to ignore the 'social' side of office life. The social nature of office work plays an important part in contributing to job satisfaction for a large number of office workers – especially for women. Case studies relating to job satisfaction and computer systems have consistently shown that *social contact* is the most important factor contributing to the overall job satisfaction of employees. As Bjørn-Anderson has emphasized:[18]

> We need contact to develop our self image and to form our values and personality. When designing information systems we must therefore make sure that we do not reduce the possibilities for social contact but rather enhance them.

New office technology poses particular dangers of social isolation to jobs which are traditionally occupied by women. As Hazel Downing stresses:[19]

> Seventy per cent of clerical workers and 98.6 per cent of secretaries and typists in the U.K. are women. The choice of paid work for working-class women is largely limited to the lowest-paid jobs in

> offices, factories, shops and industrial and private cleaning. For many women, office work stands out as perhaps the cream of the choice, offering clean, comfortable, respectable work with the possibility of promotion up through the secretarial hierarchy. . . . Perhaps most important of all, the office has traditionally been an area where the working relationship between the boss and the typist/secretary has been personal in nature

According to this argument, new office technology threatens to diminish the job content of copy typists who work in large typing pools to a much greater degree than that of personal secretaries. The status and position of personal secretaries depends not so much on her skills and her typing ability but on the status of her *boss*. The proportion of the job which relates to secretarial paper qualifications decreases the further up the secretarial hierarchy one ascends. Promotion depends not so much on high typing speeds but on 'the successful manipulation of modes of femininity which are class-specific'. In order to move up this status hierarchy, the working-class girl must display specifically middle-class modes of femininity and in particular, a tasteful dress sense, a natural telephone voice and perfect grooming. She must learn to isolate herself from her workmates and distance herself from the 'girls'. Despite the increasing specialization of tasks that has characterised larger organizations in recent years, with the subdivision of offices into separate departments such as accounts, wages, sales, customer services and post room, where each clerical worker has his or her own specific job to do, there has been a marked reluctance on the part of managers to dispense with personal secretaries. Indeed, they have multiplied. The possession of an 'office wife' has become the mark of success for many executives. Thus there is every reason to suspect that there will be significant resistance on the part of many managers to fully utilize the potential of new office technology in so far as it affects their own secretaries.

In contrast, those women who work in large typing pools can be expected to feel the full impact of new office technology. This is because the introduction of word processors effectively transfers the control which the typist has over the conventional typewriter on to the machine itself. Many word processors have functions built in to them which allow the job of the operator to be continually controlled and monitored. Word processors can indent, centre, justify margins, tabulate and print page numbers and footnotes automatically, as well as storing and retrieving standard paragraphs and letters, thus removing the skill in producing well-presented work. It is also often the case that the printing of the final documents is carried out elsewhere, so that the operator does not even see the contents of her document in its final state. Downing suggests that as secretarial work becomes gradually more fragmented, the repetitive, standardized typing will be sent to the word processing centres and

> the cream of the secretarial workforce will hang on those few jobs at the top – the two extremes of the structure becoming increasingly polarised. The loss of skills which word processing effects will make it almost impossible to bridge the gap between the two extremes.

However, Downing is at pains to emphasize that the promotion opportunities for women office workers will necessarily become more limited, and the limitations that exist already will become more visible – particularly for working-class women.

The 'social' side of the office can be severely diminished by new office technology in two ways. First, it enables typing work to be increasingly deskilled and fragmented and poses the danger of relegating large groups of typists to pools where the possibility for conversation and social contact is limited; secondly, it can also depersonalize the informal working relationships that frequently exist between the author of a document and the person responsible for typing the document.[20]

A report based on case studies of insurance companies, banks and other offices in France[21] which concentrated primarily on the changes in the quality and organization of office work brought about by the introduction of new technology on secretaries and typists, found that, in terms of work content, the functions of secretaries and typists were not significantly changed. In fact, many of the respondents found that their jobs were more interesting because they liked to work with the modern equipment and they had been relieved of the more repetitive and monotonous tasks. However, some of the copy typists complained that their jobs had been made more unsatisfactory because the machines took the discretion out of their work – largely as a result of alterations in spacing, and workplace layout. This is very much in line with the views of Hazel Downing.

Perhaps the most significant aspect of this French report lay in its identification of two stages in the development of the 'office of the future' in France. In the first stage, existing equipment and work organization would coexist with progressively advancing machines and concepts of work. At this stage no significant changes in employment levels or job content would occur. However, at the second stage the analysts believed that there would be an invasion of electronic equipment that would be increasingly interactive and at the same time, systems designers would have an 'obsession' with the rationalization of work.

Another area of the job content effects of new office technology is concerned with what may be called 'organizational ergonomics'. Whilst most experts on the social and employment effects of new technology would acknowledge that there is no technological determinism relating to its application, some of them would point out that some form of 'economic determinism' is often at work in the sense that jobs are designed to conform with the economic 'imperatives' of the technology rather than the other way round. Some of the most obvious risks to the operators are:

> specialization, e.g. one operator is left with one specialized, monotonous task instead of having any opportunity for job enlargement;
> designing jobs so that task discretion is removed from the operator, and the task is programmed and pre-planned;
> using the new technology to monitor the performance of the operator.

It is important to emphasize that if office technology advances so that the job design philosophies associated with it assume a 'Tayloristic' form, then it will serve to close more and more options and choices of good job design as time goes on. The more the division of labour between the human being and the machine shifts towards the machine, the more necessary it becomes to increase the programming and pre-planning of the particular part of the job tasks which are carried out by the operator.

What concluding remarks can we make about the effect of the new office technology on job content and work organization? In the first place, there is nothing deterministic about the new office technology; the equipment itself does not necessarily lead to the degradation of job tasks. Whilst there is some evidence from the case studies that have been carried out so far that polarization of jobs can emerge, traditional avenues of promotion can be reduced, the social nature of office work can be diminished and work can be increasingly monitored and controlled, a great deal seems to depend on the organizational changes that take place when the new technology is introduced and the motives underlying management decisions to introduce the technology in the first instance.

Secondly, nearly all the case studies are concerned with word processors and thus merely encapsulate the office technology at a very early stage of development; the future effects on office jobs must inevitably be somewhat speculative. Nevertheless, there is certainly an *a priori* case for advocating much more consultation and participation with the employees concerned whenever new office technology is about to be introduced. The potential for dehumanization of tasks certainly exists and the dangers inherent in job rationalization and specialization are all too readily apparent.

Office Technology and Job Loss

In the United States and Canada the debate on the unemployment threat stemming from new technology has not reached the same intensity that it has in Europe. This may be because many North Americans have been lulled into a false sense of security by the dramatic growth in employment over the past two decades. The growth in employment took place despite the public attention that was given to the issue of computers and possible unemployment in the 1960s. When the early computers appeared on a commercial scale in the United States there were widespread fears that the adoption of computers to take over many routine clerical functions would cause high levels of unemployment. There was even a Presidential Commission set up to consider the matter.[22]

In the event, while there was evidence that some office tasks did result in the reduction of some clerical activities, the predicted result of increased unemployment did not occur. The displacement effect was far outweighed by the increase in the volume of work generated at the level of the enterprise by new jobs created in data processing and the general increase in demand for goods and services as a result of higher economic growth rates and rising living standards. The Presidential Commission concluded that:

> the general level of the demand for goods and services is by far the most important factor in determining how many are affected, how long they stay unemployed, and how hard it is for new entrants to the labour market to find jobs. The basic fact is that technology eliminates jobs not work.

Recent estimates on employment levels from the US Department of Labor suggest that secretaries, typists and general office clerks will be the source of the largest number of new jobs in the 1980s.[23] Moreover, even with the increasing use of word processors in the United States, secretaries have been one of the fastest-growing occupational groups in the past 10 years. Against this, it has been estimated by Victor A. Vyssotsky, an executive director at Bell Laboratories, that only a *two* per cent per annum reduction of office staffs in America would be needed to displace 25 million workers by the year 2000.

In contrast to Western Europe, there have been fewer studies carried out in North America on automation and employment and most of those that do exist concentrate on the industries with the highest information intensity, like banking and insurance. With regard to office work in general, we are largely reliant on a few anecdotal accounts from business magazines.

As we have already mentioned, banking is particularly suitable to the introduction of microelectronics technology because a large proportion of banking activity consists in handling large quantities of information – much of which is in numerical form. For example, it has been estimated that about 81 per cent of the output of banking in the United States is used in providing information services.[24] As in Europe, banks in North America have increasingly computerized their activities by providing on-line computer services to customers through terminals to the central bank computer. The main impact of this development, in so far as employment is concerned, is likely to fall on the bank teller. A Canadian study of the banking industry reported that the adoption of on-line teller terminals truncated the job functions that were required to complete a banking transaction radically. Even though the traditional work of a teller had been reduced by up to one half, the report noted that this had not yet resulted in a reduction of teller personnel. Instead, the amount and type of work that they did had expanded. New job descriptions changed from the information handling traditionally associated with teller work to information marketing – where tellers were engaged in dealing with customer queries and complaints and promoting new services. Consequently, the study concluded that it was not possible to specify the net employment effects of this type of automation, although it was noted that the number of part-time workers had increased.

North American banking practice in terms of providing automatic teller machines for customers is slightly ahead of that in Europe. In 1980 there were more than 20,000 automatic tellers in use in the United States and about 100 in Canada. It has been estimated that the number of these machines will rise in the United States to 54,200 by 1985 and 71,000 by

1990. According to an American banking industry report[25] there will be a levelling off of employment growth if this projection is correct and the growth in the number of bank branches will slow down, even assuming that many teller functions cannot be automated.

In one US bank (Citicorp) automation has already been introduced into the trust department where stock transfers take place. There, the speed of transfer and quality control are of the utmost importance and the bank concerned had to process 12,000 certificates daily for more than 1,000 different issues. A move to decentralized computer work station operation enabled the number of staff to be reduced by one-third; those employees displaced were transferred to other parts of the bank's operations. The same bank was also cited in another case study[26] which involved the automation of processing letters of credit. In June 1975 it took 14 people 3 days to process a letter of credit; in 1979 one person working with a minicomputer and filing records electronically could do the same job. Within 3 months staff were reduced from 142 to 100. A small number were reclassified as professionals; others were transferred to other operations. This particular bank reduced its clerical staff from 10,000 to 6,000 during the 1970s. All of those displaced had been transferred to other parts of the bank's operations because there was an extensive re-training programme and a growth in business. It was anticipated that the bank would be able to move into higher levels of customer service such as tax counselling, budget counselling, cash management, cost accounting and portfolio management. The scope for transferring employees displaced as a result of automation of other areas of banking will be considerably reduced unless this expansion of services continues at the same rate.

The FIET prepared three reports on the impact of new technology on bank, insurance and office workers in firms that compete in international markets.[27] FIET found that major redundancies had not occurred in retail banking, although the rate of recruitment, as well as the demand for certain types of staff, had declined. There was also some evidence of an increase in the proportion of staff who were working part-time.

The American case studies on the banking and finance industry seem to suggest that while the introduction of new office technology has resulted in job displacement, other opportunities have been available to offset these job losses which can largely be traced to a substantial growth in the services provided by the institutions concerned.

Sweden is in the forefront of computerized applications, both in its manufacturing industry and in the progress it has made in introducing computers into the office environment. ADP was introduced into its social insurance system during the early 1970s. The ADP system is centrally located at Sundvall, which is about 320 kilometres north of Stockholm, with which over 1,000 viewing screen terminals are connected. The SCB estimated that the introduction of ADP up to 1975 caused the loss of 15,000 – 20,000 jobs. The Swedish Union of Insurance Employees (FF – Försäkringsanställdas Förbund), estimated in 1980 that in the National Office of Telecommunications (Televerket), 7,000 jobs would be eliminated during the 1980s.[28]

In the United Kingdom, EPOC carried out a 21-company survey to ascertain what the impact of word processors had been on the employment of women.[29] The respondent companies indicated that their objective in introducing the new equipment was to increase efficiency. The majority reported that there had been no change in the number of jobs involved. Three firms had had job losses, almost all among women, but these had been accomplished by natural wastage rather than redundancies. Although redeployments were not significant, in the cases in which they had occurred, they had led to higher-skilled jobs. Six companies reported that employment had increased: this was thought to be in response to an increase in business. However, one employer said that 'without these products there would have been some job loss but the current prospects are for job expansion'.[30]

Similar findings emerged from Emma Bird's study for the Equal Opportunities Commission.[31] In three of the nine companies studied there was a net reduction in jobs reported among secretaries and typists. In one case the job loss did not result solely from the use of word processors but as part of a rationalization of secretaries within the organization. In the other two companies reporting job loss, large productivity gains were realized in terms of text turn-around time. In both cases the use of word processors was highly centralized in pools.

In fact, there appear to be four possible structures which can be linked with the introduction of word processors into an organization. One is a centrally administered system where all correspondence secretaries are located at the word processing centre and all typing within the organization is sent there. An alternative is to have satellite centres with typists trained in specialist typing skills and housing both administrative and correspondence secretaries. A third structure would require a back-up centre as an overload system for traditional secretaries. A fourth structure is to have a decentralized.system where word processing facilities are located in the normal departments and less division takes place between the work of administrative and correspondence secretaries. The *less centralized* the system, the less polarization of skills will take place.[32]

One thing that does emerge from the literature is that whilst there is ample scope for massive job displacement as the technology develops and is widely diffused, much depends on the way it is introduced and how work is reorganized as a result.

The Impact of New Office Technology on Women's Employment

During the past two decades women have entered the labour market in increasing numbers. The majority of them have found employment in the tertiary sector of the economy and in non-manual occupations. For example, 72.3 per cent of women in the UK were employed in the services sector in 1977; in the US, Canada, and the nine EEC countries the figures were 80.5, 81.0, and 66.3 per cent respectively. This long-term trend of a growing share of female employment which was already evident between 1960 and 1970 became more pronounced between 1973 and 1980. The number of women in civil employment in the OECD countries rose from 34.5 per cent in 1960 to 36.2 per cent in 1973 and 38.6 per cent in 1980. In 1980 it was more than 40 per cent in the US, the UK and the Nordic countries, but less than 30 per cent in most southern European countries and in Ireland.[33]

As women have entered the labour market, certain activities in the service sector and occupations have become increasingly 'feminized'. One of the important characteristics of the labour markets in advanced industrialized countries is the marked segregation by sex. Women are concentrated in a limited range of occupations and are most likely to be found working in relatively less-skilled and lower-paid jobs than their male counterparts. Clearly, this concentration has important implications for those areas where new technology is expected to have the greatest impact; many of the occupations to which women are attracted are precisely those which the new information technology will transform and/or eliminate. Many female office workers may find themselves having to work at home (see chapter 1).

In particular, microelectronic technology will dramatically affect those occupations which can be described as information-related (i.e. those concerned with the creation, processing, storage and transmission of information). A recent OECD study provides an inventory of 'information occupations' which are listed under the ILO International Standard of Occupations as follows:

1. The *information producers* create new information or package existing information into a form appropriate to a particular recipient (engineers, agronomists, statisticians, market specialists, insurance agents, song and lyric writers, for example, but also all kinds of consultants including doctors and system analysts).
2. The *information processors* receive and respond to information inputs (administrative, managerial, supervisory and clerical staff, etc.).
3. The *information distributors* primarily convey information from the initiator to the recipient (education and communication workers, journalists).
4. *Information infrastructure personnel* install, operate and repair the machines and technologies used to support information activities (telex and office machine operators, repair and maintenance staff, postal and telecommunications workers).

For the nine countries which participated in the OECD study it was found that, on average, the share of information occupations in the labour force has increased since 1950 by nearly 3 per cent in each 5-year period. In 1975 these occupations accounted for more than one-third of the labour force.

Of course, the information sector of the economy is not homogeneous. Within it are skilled occupations – those who create, analyse, co-ordinate and interpret information – and the less skilled (those who manipulate or handle information). Women tend to be concentrated in the latter category of activities in occupations such as secretaries, cashiers, stenographers, bookkeepers, filing clerks, or typists. In contrast, the upper echelons of information occupations are dominated by men who constitute the vast majority of senior and middle management and professional workers. Information technology applications are primarily conceived of as a tool to assist the latter group in their decision-making, analysis and communication by speeding up and broadening the flow of information to and from them. Labour statistics data from the OECD countries show that the information-handling occupations are predominantly *female* and the degree of sex segregation becomes evident. For example, in the United Kingdom one-third of women work in offices, and of these more than 90 per cent are employed in routine clerical jobs. The top three female occupations are clerical, cashier and typist jobs.[34]

What alternative opportunities will be open to those women who are displaced by the new technology? Not only are there likely to be fewer jobs to go to as a result of technological change, but also women generally have limited access to such alternative job opportunities because of their domestic and family commitments. They are less mobile geographically and the longer they are out of the labour force the more likely it will be that the need for their traditional skills of typing, shorthand, etc. will have declined. Older women in particular will find it more difficult to adjust to the new office technology.

Even for those women who remain in office employment, many of them may well face increasing specialization and fragmentation of tasks which are continually monitored and controlled. Given such a scenario, work would be organized around manual rather than mental tasks and jobs would become more structured and depersonalized. Case studies so far show that there is a disturbing tendency for jobs to *change gender* when new office technology alters job design to require a greater degree of technical expertise.

Economists have noted the paradox that while the number of married women in paid employment has dramatically increased, sexual inequality in the labour force has largely remained static. If more and more female office jobs can be done at home (with lower pay) then there is a danger that these sexual inequalities could become more pronounced.

Another feature of the information sector is that it is characterized by low levels of unionization – particularly among women – which means that they are likely to be in an unfavourable bargaining position to negotiate about the introduction of the new technology and to procure a share of the gains of technological change. Although we must be cautious in predicting the range of new jobs that will be created by the spread of the new office technology, it does seem that computer-related occupations such as programmers, system analysts, data management and electrical engineering will grow in numbers – the vast majority of which are held by *men*. Even when women work in such higher-level occupations, in comparison to their male counterparts women tend to be concentrated at the lower end of the skill spectrum.

It is well known that women tend to be located in certain occupations as a result of their education. Few girls specialize in technical or mathematical subjects at school; the majority choose courses in the arts, humanities and languages rather than science, mathematics or other technical subjects. In the vocational courses that are open to both men and women, a majority of women opt for courses in general office employment and other services whereas men choose apprenticeships in technical skills or industrial trades.

The rapid spread of the new technology makes it all the more urgent that women acquire new skills in order to broaden their occupational base. As Werneke points out:[35]

> Basic change must begin in school where girls must be influenced, both within and from outside, to participate more fully in technical subjects that will give them the analytical abilities necessary for more promising jobs in the future. This influence should be extended to post-school training where women should be encouraged to undertake broad-based training in business and the new information systems. . . . Merely replacing a typewriter with a word processor in the classroom is not going to provide women with the training they will need as electronic systems are introduced in the office. The concept of training must move from providing task-specific skills to giving a broader understanding of an organization as a system and must lay the basis for the continuous retraining that will be necessary over a career.

In conclusion, women are very vulnerable to the impact of new office technology. It is somewhat ironic that the growth of office employment in the Western economies during the 1960s and 1970s opened up a wide range of job opportunities for women to enter the labour market and yet it is precisely those jobs that are most vulnerable to new technology. There may also be a certain irony in the fact that with so much attention paid in the 1970s to reducing the alienation of the blue-collar by work redesign and job enrichment (see chapter 3), new office technology could reproduce the same routinized, highly structured, controlled and depersonalized work.

Health and Safety Aspects of Office Employment

A great deal of attention has been paid to the health and safety aspects of new office technology, particularly on the increase in stress which many claim is brought about by the increased pace of work and on the health hazards of VDUs. There is evidence that stress is associated with intensive use of computer technology, particularly when the pace of work is dictated by the machine rather than by the operator. In addition, it has also been shown that long hours at the terminals of VDUs can cause eye strain, headaches, back fatigue and migraine and these problems can be compounded by inadequate lighting, poor ventilation, uncomfortable chairs and high noise levels. Many of these problems are being alleviated by the provision of better monitor VDU screens, improved ergonomic design of workplaces, the use of regular eye tests and the provision of frequent rest breaks from work at the terminal. There is little or no evidence to suggest that VDU radiation emission levels can be dangerous.

Conclusions

One thing that does emerge from this chapter is that technology alone cannot guarantee improved performance and higher morale in an office. We are still in the embryonic stage of new office technology, and many changes in work organization, work patterns, skills and employment levels are yet to come – particularly when large organizations link up their automatic data processing to telecommunication systems.

In the next 20 years or so it seems likely that automation will enter the office environment in a pervasive way. Those clerical workers who remain in office employment will sit in front of VDU terminals (either at home or within the employer's premises) as they handle inquiries from customers, prepare memos and reports, and communicate with their co-workers. The real danger is that office work will become 'Taylorized' and proletarianized as the link between office machines and telecommunications enables workers to be subject to increasing management control and monitoring.

It is unlikely that many offices will become 'paperless'. Rather, computer and telecommunications technologies will be more widely used to replace and to complement other information media. Office environments should generally improve as people become more aware of the discontent that can arise from poor job design. The diffusion process of office technology will be very uneven and its speed will depend on the size of the organization, the diversity of tasks, market factors, and managerial strategy generally.

The office sector is a crucial area for changes in work organization; it will be the sector of the economy which is most affected by information technology – at least in the foreseeable future.

3

On the Shop Floor

In chapter 1 when we looked at the impact of new technology on the enterprise we noted how microelectronic technology would affect manufacturing in two main ways. In the first place, microelectronics would increasingly be incorporated into new or considerably modified products. Secondly, it would dramatically transform the manufacturing of traditional products by the use of CAD and CAM, FMS, robots and other automated devices. In this chapter we will consider the implications of these two factors in much more detail, concentrating particularly on the effect they will have on job design, skills and employment.

The Evolution of Automation in Manufacturing

All production processes essentially consist of three different functional activities: the *transformation* of work-pieces; the *transfer* of work-pieces between work-stations; and the *co-ordination and control* of this transformation and transfer. It is possible for all or some of these activities to be be mechanized or carried out manually.

Three phases of mechanization can be identified which can be linked with these functional activities. The first of these phases – *primary mechanization* – started around the middle of the nineteenth century and continued until the beginning of the twentieth century. Here, the main emphasis was on the use of power-driven decentralized machinery to accomplish transformation tasks. The second phase – *secondary mechanization* – involved the use of machinery to accomplish transfer tasks, and ran from about the beginning of the first world war to the 1950s. Of course, the above outline is highly generalized and by no means adequately describes what has been a complex series of developments from one manufacturing industry to another. For example, as Littler argues[1] primary mechanization is still continuing but is still likely to occur together with secondary and tertiary mechanization. Secondary mechanization has already advanced significantly, but further diffusion is restricted by product markets. As for *tertiary mechanization*, by 1975 it had already been adopted in some industries and since then it has become increasingly more flexible.

F. W. Taylor and Scientific Management

It is important to consider the history of mechanization of manufacturing alongside the developments that took place in management ideology. Without doubt, the most significant influence on management thought with respect to automation and control of production is that of *'Scientific Management'* or *'Taylorism'*.[2] At the end of the nineteenth century, workshop administration in manufacturing industry in America and Europe was still largely in the hands of foremen and skilled workers who, in addition to performing the physical tasks of production, decided how jobs were to be done, how the labour force was to be organized and supervised, and in many cases also decided who was to be hired. Frederick Taylor advocated that the administration of the work place should be transformed so as to ensure greater management control, profitability and efficiency of production.

Taylorism essentially consisted of three main principles. Firstly, it involved the maximum division of labour so that production processes were systematically analysed and broken down into their component parts, in order to simplify each worker's job and reduce it as much as possible to a number of simple tasks. Greater specialization would therefore lead to greater efficiency, while the deskilling that followed this task simplification would allow cheaper unskilled labour to be hired. One of the main objectives of Taylorism was to ensure a greater division of labour which would in turn remove the planning, organizing and hiring functions from the shop floor.

Management organization was also to be increasingly specialized, which would help to ensure full managerial control of the work place. Management control would also be supplemented by a cost accounting system based on a time and motion procedure, which would provide managers with the information they needed in their new roles as controllers of the work place.

In sum, scientific management proposed two major transformations of manufacturing industry simultaneously: the removal of manual skills and autonomy in decision-making by shop-floor workers; and the establishment of a number of 'scientific principles' of management as distinct from ownership.

Scientific management principles had some success in the United States prior to the first world war and later in Europe, including the Soviet Union under Lenin. However, it was by no means universally adopted. Indeed, there was a great deal of resistance from workers (and even from some managers) to the pure forms of Taylorism. Moreover, there are limits to the division of labour which are implied by Taylorism. The division of labour depends on a certain level of output which in turn depends on a mass market. If a certain piece of work involves several operations it is not economical to employ a specialized worker for each work task involved in manufacture where the total volume of output only requires the time of one person.

Henry Ford and the Assembly Line

The link between mass markets and the division of labour enabled Henry Ford to maintain a mass market for automobiles between 1908 and 1929, when the last of over 15 million model T cars rolled off the assembly line. Ford employed a system of production control to serve this mass market based on a form of Taylorism, except that he extended Taylorism by transferring skills from workers to specialized machines on an assembly line so that work could be brought *to the worker*. The pace of work was therefore controlled mechanically and not by workers or supervisors.

Associated with the new fixed-speed moving assembly lines was an accelerated division of labour and short task-cycle times. Unfortunately, Ford's production methods produced high levels of absenteeism and labour turnover. To remedy this, Ford adopted a new control technique which marks off Taylorism from 'Fordism'. Whereas according to one of Taylor's close associates Taylor did not 'care a hoot what became of the workman after he left the factory at night, so long as he was able to show up the next morning in a fit condition for a hard day's toil'[3] Ford tried to remould working-class culture outside the factory gates in an attempt to induce greater industrial discipline and responsibility. He introduced the principle of the 'five dollar day'. This was a wage package which guaranteed high rates of pay. However, in order to qualify for this wage package, workers had to have been continuously employed for 6 months, be aged over 21, have satisfactory personal habits at home (cleanliness and prudence) and not to consume alcohol or tobacco. These standards were continually checked by a newly-established 'Sociological Department'(!). The diffusion of Taylorist and Fordist ideas was very uneven throughout the West, but in general by the mid-1930s, Taylorite techniques had spread across Europe whilst Fordism and the moving assembly line had penetrated the largest car firms and spread to other industries such as electrical engineering.

There are several reasons why both of these philosophies of managerial control were not applied universally. There are limits which serve to restrict their application throughout manufacturing industries. Economic factors limit the complete decomposition of work tasks because such decomposition depends on the velocity of throughput. There are also technical limits in the sense that the division of labour can only be carried so far – beyond which further decomposition of tasks must await a transformation of machining technology. Any attempt to extend Taylorism and Fordism also necessitates an extension of co-ordination and control procedures which carry high costs in terms of production planning, supervision and monitoring. Finally, such systems of production organization are limited by the extent to which workers themselves are prepared to comply with a system of managerial control which is is founded on monotony, boredom and is devoid of any kind of job autonomy. The costs of ensuring worker compliance with such systems are much higher during periods of full employment. This is one of the reasons why job enrichment schemes were so popular in managerial circles during the late 1960s and early 1970s.

Job Redesign, the Quality of Working Life, Group Technology and Quality Circles

The influence of Taylorism and Fordism can be seen in many modern systems of production organization. There is a tendency for job tasks to be designed to fit in with the technology rather than the other way round

> . . . there is still a strong commitment to the proposition that meeting the requirements of the technology [process, equipment] will yield superior job performance, measured by organisationally

> relevant criteria, and a deep-seated conviction that the same performance will not be achieved if technological requirements are not given exclusive consideration.[4]

The late 1960s and early 1970s saw the emergence of a new job re-design movement which was largely based on industrial psychology and the 'Quality of Working Life' (QWL) movement. Two particular reports were symbolic of this movement; the *Work in America*[5] report of 1973 and the *On the Quality of Working Life*[6] report in Britain in the same year.

These reports, in addition to the consultants who were involved in advising firms about the best way to reorganize their production systems, advocated five principles of 'good' job design which were a radical departure from Taylorism:[7]

(a) The principle of closure – the scope of the job should include all the tasks necessary to complete a product or process. Theoretically, the predicted result is that work acquires an intrinsic meaning and people can feel a sense of achievement.
(b) Incorporation of control and monitoring tasks. Jobs should be designed so that a large number of inspectors are not required. The individual worker, or the work team, assumes responsibility for quality and reliability.
(c) Task variety, i.e. an increase in the range of tasks. This implies a principle of comprehensiveness, which means that workers should understand the general principles of a range of tasks so that job rotation is possible.
(d) Self-regulation of the speed of work and some choice over work methods and work sequence.
(e) A job structure that permits some social interaction and perhaps co-operation among workers.

Despite the changes in job design which were advocated by the disciples of the QWL movement, apart from a few isolated examples – particularly in Scandinavia – job design remained tied in most manufacturing enterprises to traditional Taylorist principles. Nevertheless, as Kelly points out,[8] the limited number of changes that have taken place have been of three main types: reorganization of assembly lines, group technology and job enrichment.

The reorganization of assembly lines has largely been concentrated in the consumer industries – particularly in electrical appliances. Here, product market pressures associated with increased product variation in the face of stiffer competition has forced companies to create more flexible work forms able to accommodate more rapid product changes without creating an entirely new line.[9] Some companies redesigned their work so that buffer inventories of partly processed goods were introduced between work stations and all automatic operations were grouped together. This enabled workers to be decoupled from the machine pace by these buffers. Although the line itself was not so highly mechanized, the jobs were slightly enlarged and the task cycles were longer. The important point was that although the initial investment costs of the new line were higher, capacity utilization increased because of greater flexibility. The new work arrangements could adapt to product changes much faster and thus reduce overall unit costs. As far as assembly lines were concerned, changes in job design still generally resulted in one-man work stations, but more co-operation between workers was possible.

Group technology was also a popular policy panacea among avant-garde management consultants in the late 1960s and throughout the 1970s. It was extensively applied in car manufacture in the famous Volvo experiments in Sweden and also at Volkswagen in West Germany. The basis of group technology is the realization on the part of employers of the value of work groups and team-work in solving production problems. Instead of the traditional fixed-speed assembly line with short task cycles, Volkswagen began experiments which enabled workers to be de-coupled from the assembly line to work in groups of seven workers, with two teams on each shift. Within the groups four men worked on assembly, two did testing and one man was in charge of materials.

According to Jenkins,[10] the entire group was de-coupled from the machine-paced line but had to meet a quota of seven engines per team per day. The workers received special training so that they could do all the team jobs and were free to rotate job assignments as they wished. Each group had a team leader (*Gruppensprecher*) who was responsible for liaison with management, and foremen were eliminated.

The experiment at Volkswagen highlighted two major problems which beset many such experiments in group technology and job enrichment during the 1970s. Employers saw the reorganization of work as a means of using the groups to undercut trade union influence within the plant, and the trade unions responded by opposing the use of informal work group leaders as potential usurpers of union influence. The eventual outcome was that the team leaders were converted into shop stewards, and foremen were brought back to oversee the groups. The reorganization also produced fresh union wage claims which centred on the question of how the specially trained team workers fitted in to the wage and skill hierarchy, given that they were multi-skilled and had wider responsibilities. The Volkswagen experiment ended in 1978; it was considered by management to be much too costly.

The experience of group technology and job enrichment schemes on assembly lines illustrates the inherent tension between participation and the need for management control. Once workers have experience of working in such officially sponsored groups and show competence in putting forward valuable ideas for work layout changes, etc., they may well believe that such participation can be extended to include matters such as work assignments, the allocation of rewards, and even the selection of team leaders.

However, the introduction of group technology in batch production has been much more complex. When group technology was first popularized in the early 1960s it was essentially a technical term – referring to a new layout of production based on grouping together all the machines necessary to complete a particular type of component. This in turn was based on a classification of components, which were standardized as far as possible into 'families'. It improved machine utilization and it speeded up the throughput of work by simplifying work flow. In doing so, however, group technology created a tendency for the development of flexible work groups based on generalized, skilled machinists with a lack of rigid job boundaries. In so far as charge-hands and foremen were eliminated, it reduced control and co-ordination costs to the employer. Such groups also increased worker adaptability and enabled the team to cope with any absences on the part of its members. Group technology spread to around 10 per cent of batch engineering firms in Britain by the mid-1970s but then the whole process came to an abrupt halt. This was because the information burden of setting up a reliable group technology production system was too great; there were too many variables and there was too much unpredictability.

The increasing possibilities which are opened up for fuller automation of batch production by the use of CAD/CAM systems and flexible machining centres in the 1980s and beyond will lead to a major acceleration of new work organization systems. This topic receives much more detailed attention later in this chapter.

Many people associate *quality circles* or QCs as being of Japanese origin, but in fact they originated in the United States. Quality circles are small groups of workers of between 5 and 20 people which meet regularly to study production problems and try to offer solutions for improvement and efficiency. They are usually led by a foreman or senior worker and are seen as a means of stimulating motivation and involvement on the shop floor. Such groups are essentially different from earlier human relations ideas because they involve systematic training of shop-floor workers and access to the necessary technical information in order to solve problems. QCs were transplanted to Japan during the 1960s and quickly gained in popularity. They were re-cycled and later taken up by several companies in the West during the 1970s – surrounded by the aura of Japanese productivity and economic success.

By 1982 it was estimated that about 450 corporations in the USA and Canada made extensive use of QCs, which can be seen as a break from the implicit practices and assumptions that lie behind Taylorism – in the sense that there is an assumption that workers will devote time and commitment to improving working practices. Moreover, management acknowledges that workers have the expertise (given adequate training and resources by management) to contribute towards the success of the company in which they work.

It is much too early to evaluate the success or otherwise of the adoption in the West of QC ideas from Japan. The Ford Motor Corporation sought to implement the system in all of its 25 manufacturing and assembly plants in Western Europe as part of its 'After Japan' programme in an attempt to improve the quality of its products, reduce scrap, and encourage worker involvement.

There were two significant differences between the Japanese system of quality circles in so far as Ford was concerned. First, the QC groups of 8 – 15 people included a supervisor plus a representative from the quality control department – a legacy from Taylorism in the sense that there was still the belief on the part of management that quality control and production planning problems should be divorced from the shop floor. The second difference from the Japanese situation is that in Japan the labour force is all-Japanese and well-educated, whilst in Britain the labour force consists of a mixture of workers of West Indian, Asian and southern European origin as well as British workers. The attempt to introduce quality circles into British Ford plants aroused the hostility of the Transport and General Workers' Union who saw the proposals as cutting across existing union structures and as a means of circumventing shop steward representation on the shop floor. So far quality circles have not been introduced into British Ford plants, but they have spread to other Ford plants in continental Europe.

The short summary given above of the evolution of automation on the factory floor shows that management strategies to establish greater control over production have evolved in line with technological innovation and changes in the political and economic context of the wider society. Whilst Taylorism has by no means been universally applied at least in its pure form, it still figures prominently in management thinking. The advent of information technology-based systems of production now provides an opportunity for management to increase their control over production systems either by eliminating workers altogether or by using unemployment and the recession as a means of enhancing their bargaining power.

This provides a convenient departure point for considering several questions concerned with the new information technology. What is the significance of robots as one of the new forms of factory automation? Does new technology by its very nature lead to the erosion of workers' skills? Are we likely to see the completely unmanned factory on a widespread scale? What are the prospects for new forms of job design? What will be the reaction of workers on the shop floor to new forms of computerized manufacture?

The Use of Robots in the Factory

Robotics, a term derived from the Czech word 'robota', meaning work, has now come into common usage to describe the science concerned with programmable machines designed to carry out a variety of tasks. The world population of programmable industrial robots in 1980 was estimated at around 15,000 units, twice the 1978 total; some 40 per cent of these were located in Japan, 25 per cent in the United States and the rest in Western Europe. These are probably over-estimates, however, since the Japanese definition of robots is somewhat less strict than the American version, which means that the proportion of Japanese robots also appears to be exaggerated. Currently, most robots are installed in the automobile industry where they are used for welding and surface treatment. In future they will be used more and more for assembly operations, providing that they can be equipped with adequate sensory devices.

According to the British Robot Association, in 1981 the usage of robots could be broken down into six main broad categories of application: 28 per cent were employed in loading/unloading; 23 per cent in manipulation and process; 36 per cent in joining; 12 per cent in surface treatment; 2 per cent in assembly and 18 per cent in an other/unknown category.

There is a great deal of misunderstanding about what robots actually are. Many people think of robots in terms of the similarities they have with human characteristics, but in fact there is a large variety of robotic devices, each with a different range of operating characteristics. At one extreme a simple pick-and-place device will have a restricted range of movements, with very severe limitations on the extent to which it can be programmed. Such devices will be programmed by means of screw-adjusted mechanical stops for setting the range of movements, and electromechanical or pneumatic switches for selecting the desired switches which activate particular operations in its sequence – which may only consist of a few steps. At the other extreme, much more sophisticated devices have a much greater range of movements and they can be programmed to carry out very complex and variable sequences of movements.

It is important to recognize that robots only form (part) of an overall system and they operate in interface with other machines, which themselves are linked in to the information flow that directs the technology. However, robotic technology is developing rapidly and there is every indication that before long we will see the development of integrated manufacturing systems comprising conventional automation equipment, together with computer numerically controlled machines which can be serviced and physically integrated by industrial robots under the overall control of a central computer.

A great deal of interest and concern has centred on the job displacement effects of the introduction of robots into manufacturing. Perhaps the most dire prediction was published in the *US News and World Report* in 1982.[11] This study, which was carried out by Carnegie-Mellon University predicted that: '. . . by the year 2000, robots will supplant 3 million factory workers and by 2025, could be handling virtually all manufacturing chores'.

A number of sectoral studies have also been carried out in Europe on the employment impact of robots which are also pessimistic, albeit in varying degrees (see Box 3.1).

BOX 3.1

During the preparation of the Eighth French Plan (1981–5), it was estimated that the introduction of robots would lead to a loss of about 30,000 jobs, which was not even 1 per cent of the French labour force in 1980. Forecasts in the Federal Republic of Germany are more sombre. Assuming that the utilization of robots will increase from a mere 2 per cent in 1980 to 60 per cent in 1990, the result would be 200,000 lost industrial jobs – about 6 per cent of industrial employment. The trouble with these sectoral studies, of course, is the fact that they are restricted to a particular sector, and they therefore do not take into account the employment effects in other parts of the economy, which are not necessarily negative.

Other German studies tend to be pessimistic about the employment effect of introducing robots into manufacturing. According to a study by the Commerzbank, about half the 1.2 million assembly-line jobs will be threatened by the second generation of 'intelligent' robots before the 1990s. The same study estimated that every robot employed in industry at the present time replaces three workers on average. The new robots, equipped with sight and a finer sense of touch, will each replace five workers or even 10 in certain assembly jobs. Volkswagen estimates that second-generation robots will do 60 per cent of all work in the car industry during the 1990s. Even the OECD believes that that even by 1987 15 per cent of assembly jobs across all American industry will be carried out by robots.

Source: World Labour Report, International Labour Office, Geneva, 1984.

What effects do robots have on job design and skill levels? So far, robots have been introduced into areas of manufacturing which consist of hazardous, dirty or monotonous work – such as paint-spraying and welding. In other words, robots have been taking over jobs which few people want anyway. In Japan, where the use of robots has been the most extensive, many robots have been used to fill a gap caused by a serious shortage of skilled welders.

What will jobs be like in the highly robotized plants of the not too distant future? Although it is too early to come to any firm conclusions on the effect of robotic installation on job design and skill levels, it does seem that the division of labour and the content of jobs will change dramatically as robots become more sophisticated over the next few years. There will be a greater distance between work stations and less contact between workers. Whilst the conditions of work will probably be safer with the worker more separated from the work process, higher production rates may reintroduce some hazards in the high-speed plant where a small failure in the system may compound to frightening proportions in a matter of seconds.

It is also likely that workers will have a responsibility for a larger span of the production equipment because there will be greater system integration and fewer workers manning the production equipment. Many workers will merely carry out a monitoring role – patrolling and inspecting a system when all appears to be functioning properly. However, he or she will expected to react sharply to an incipient crisis and take corrective action when anything goes wrong. This may involve knowledge and competence of a depth essential to handle a highly integrated system of machines. Thus more jobs will be integrated and old boundaries between tasks will be wiped out as jobs are combined and enlarged. There is also the possibility that as robots become more and more sophisticated and complex, many workers will find their jobs less demanding and devoid of any interest or satisfaction.

Social contact between workers will also be severely reduced, because there will be a greater physical distance between work stations. The highly robotized plants of tomorrow will no longer be dependent upon the establishment of work-teams with a strong sense of cohesiveness, and isolation from other workers will be a common feature in such plants.

Advanced robotized plants are also likely to demand different kinds of supervisory skills. The traditional human relations skills of the supervisor will become less and less important as a premium is placed on technical competence. There will be fewer levels of supervision and fewer workers. The need to ensure the maximum utilization of costly equipment and the necessity of avoiding shutdowns will lead to much closer supervision. Production schedules are also likely to be much more demanding. Supervisors will also find that as their traditional human relations skills are not so important they will become increasingly dependent on the skills and knowledge of individual workers. As Katzman stresses:[12]

> Since the intricate technology will not allow him to become an expert, the supervisor will need to be certain that his employees are well trained and that they are functioning at their peak so as to match the speed of the plant.

Just as the design and structure of jobs in highly robotized plants will change radically, so too will career patterns, channels of promotion and job security. There will be fewer workers, fewer job classifications, fewer levels of supervision and increasingly complex technical work environments. Many workers will find that that the promotional ladders that they planned to climb will be shortened or shattered.

What levels of pay can these workers in advanced robotic plants expect to receive? According to Katzman:[13]

> Work in the highly robotized plants will be generally rewarded with higher pay. While there will be little up-grading of jobs in the new plants, workers will be paid a little more per hour than in less automated plants. With self-regulating mechanisms controlling robots and built-in rates of production, there will be little the worker can do to augment quantity or quality of a product. This will mean that individual or small group incentive systems will no longer be useful.

Despite the attention that robots receive from the media, it is important to reiterate that robots are only one component of a whole series of technological devices which are based on new technology. The impact that they will have on employment, skill levels and the design of jobs cannot be considered in isolation from other forms of automation in manufacturing.

Microelectronics Applications in Manufacturing

Much of the discussion so far has been largely concerned with automation of assembly lines. However, most metal goods are made in small batches of a few hundreds at a time, for which automation has so far been prohibitively expensive. Only manufacturers producing items in thousands or millions have been able to afford large-scale investment in automated specialized machinery. With the advent of microelectronics, it is now possible for automation to spread beyond the confines of mass production to *batch production systems*. This is a significant development because small batch production constitutes around 70 per cent of world-wide manufacturing industry.

Until the advent of sophisticated electronics and computers after the second world war, the automation of production processes could generally be described as 'hard' automation, where automation involved the use of stops, cams, and so on, so that human intervention was not required. This type of machinery has its origins in the nineteenth century and still continues to be important today. However, its use is limited to mass and large batch production, since once set up it is normally a very costly and time-consuming process to adapt the machinery to even slight modifications in the product. Indeed, product modifications may even mean the scrapping of whole sets of machinery which are of no use for any other purpose. Thus many sectors of the engineering industry, where small and medium-sized batches predominate, have found little use for those conventional automatic machines, and by far the majority of batch production processes have so far remained manually controlled.

As was pointed out in chapter 1, the automation of batch production is on the increase, and automation even for very small batches or 'one-offs' (single components) is rapidly becoming economically viable. This is because electronic and computer-controlled machinery enables control to be exercised by the programmable software of the machine's electronic control system. Essentially, computerized automation is made possible by the flexibility which comes with programmable control – reprogramming and resetting is easy compared with adapting hardware. The increasing sophistication and decreasing costs of electronic control systems due to the availability of microelectronics has enormously extended the range of feasible applications. Virtually every sector of manufacturing industry which uses multipurpose machinery is either already under competitive pressure, or likely soon to come under pressure to make use of the new control devices.[14]

The development and application of NC for machine tools began in the 1940s and 1950s and spread rapidly throughout the engineering industry. By 1978, NC machinery represented about 1.5 per cent of all machine tools in Britain.[15]

With NC, information on the size and shape of a component is fed into a control device by a paper or magnetic tape. This information is then passed by a series of electromechanical devices which can run the machine. This has several obvious advantages: better flexibility; pre-coding of jobs leads to greater machine tool usage; and tapes can be filed and recalled whenever needed to cater for minor design modifications. The increased costs of depreciation, maintenance and programming of NC machines can usually be offset by higher output, assuming utilization is kept high.[16]

More recently, CNC has been developed, where a powerful mini or micro-computer is installed on the machine tool itself. With this system, the programme is read only once for a whole batch of components, the programme being stored in the computer on the machine itself, thus reducing total tape preparation time and increasing utilization. CNC has also facilitated the development of DNC systems where several machine tools are controlled simultaneously by one central computer. However, while CNC has now virtually replaced NC, DNC is still mainly in the experimental phase, and the cost of these systems is likely to be prohibitive for some time, although some experts predict that such systems will become economically viable within the next few years. A related technology, FMS, consists of three basic elements: a set of machinery stations, a transfer mechanism, and a central computer which oversees the entire operation. Again, this technology is still in the experimental stage (there are about 100 FMS systems world-wide including the famous Yamazaki machine tool plant in Japan).

At the same time as machinery on the shop floor is automated, so too is the design and drawing of engineering parts in the office by means of CAD. Thus CAD/CAM has evolved, and in the near future it will be possible to produce programmed tapes directly from some CAD facilities. It is only a matter of time before CAD/CAM can be directly linked to the whole production system incorporating robotics, thus leading to a completely automated factory.[17]

Clearly, such technological developments have dramatic work and employment implications. Before considering what these are it is worth discussing whether or not the completely automated factory is a distinct possibility – at least in the foreseeable future.

Towards the Automated Factory?

The idea of a completely automated factory as the culmination of technical ingenuity has fascinated engineers and social visionaries (albeit for very different reasons) for a considerable period of time. However, the notion of a completely automated (unmanned) factory is seriously misleading; such factories will still need people, but not many. There will still be a need for unskilled labour – at least in the short term – to load and unload parts and to clean and clear the equipment. There will also be some semi-skilled tasks of changing tools where automatic changing is too difficult. Factories in the future will also require three levels of highly trained engineers: planners at the overall programme level, computer operators at the computer production level and technicians at the workshop level.

Moreover, the impact of FMS technologies will not be felt overnight. As mentioned above, there are still only around 100 FMS systems in the world and these plants are by no means completely 'flexible'. They can turn out a variety of products by reprogramming the computer, but only within the same family of items. There is a vast difference between a highly automated plant staffed by a group of monitors and machine-minders and a factory with *no-one* on the shop floor. Perhaps the reality in the foreseeable future will lie somewhere in between these two extremes (see Box 3.2), although the impact on work and employment will be immense. Most importantly, the labour content of the production process will be drastically reduced.

BOX 3.2

The authors of a recent Anglo-German study point out that the fascination of the automated factory is founded on past experience with the expansion of homogeneous mass markets, and stable demand for specialized products over time. It is certainly true to say that in continuous flow production processes, where the transport and feeding of throughput is less problematical, such as refineries or ethylene plants, there has been a 'flight' away from the shop floor with production operatives becoming rarer and less involved in the production process.

However, the authors claim that we have now entered a new phase of technological advance which is characterized by more differentiated and shifting market patterns which always bring about the need for direct human intervention because standard solutions are too costly, complex, and liable to be unworkable. The present trend of technological advance in batch-manufacturing appears to combine productivity increases and greater application of microelectronic technology with the continued presence of 'directly intervening' production operators on the shop floor. Labour on the shop floor is not necessarily 'crowded out', and the present phase of CNC application is marked by greater interchange of personnel between production and planning departments as well as a drift of planning and programming functions on to the shop floor. A more mechanized factory may be more capital-intensive without necessarily 'crowding out' direct labour, and the work content of operators can evolve further towards planning, programming and setting functions.

Source: *Microelectronics and Manpower in Manufacturing: applications of computer numerical control in Great Britain and West Germany,* A. Sorge, M. Warner, G. Hartmann and I. Nicholas, Research Report, Wissenschaftszcentrum, Berlin, 1981.

One implication of the Anglo-German study outlined in Box 3.2 is that designers of microelectronic application systems for manufacturing industry and the engineers and managers who operate such systems *do* have the scope to maintain and extend skills; as already emphasized, there is nothing deterministic about the new technology and the degree of discretion available to managers, engineers and designers is sufficient to allow for considerable variations in the technical and social organization of work.

Microelectronics in Batch Manufacture: The Shop Floor Reaction

A number of studies have been carried out on the effect of the use of microelectronics in small batch manufacture. Many of these studies highlight the choice that exists between using the introduction of the new technology as an opportunity to build upon the skill and experience of workers or using the new technology as a means of degrading such skills in a Tayloristic fashion, replacing them with programmes and systems devised by an array of 'experts' (see Box 3.3).

BOX 3.3

Two contrasting examples illustrating the choices available to managers in job design were reported by Wilkinson. Here, an engineering company decided to automate a plating line, where components were dropped into various chemical solutions to coat them with zinc or aluminium. Under the old system three men worked on each plating line and loaded parts onto a system of fixtures and controlled the fixtures as the parts were dipped into vats of different chemical solutions. The platers also unloaded the components at the end of the process.

The new plating system was computer-controlled and the carriage running along the rails over the various chemical solutions could be preprogrammed to drop or lift containers of jigs (holding the chemical components) into and out of the various chemical solutions. The new automatic process only required two men simply to set the equipment and to unload and load as necessary.

Management saw the work-force as 'lazy, unreliable, stupid and untrustworthy' and installed the control panel well away from the plating line so that only the works manager, general manager, and the chief maintenance engineer had access to it. The plating line had to be fitted with a manual override to allow the platers to intervene if an emergency arose as a result of a faulty computer program.

Initially, the new system broke down because of computer program faults. Yet even after these programs had been proved, workers continued to use the manual override to control the pace of the line whenever they had the opportunity. The platers still believed that their observational judgements led to a better product being produced. Thus new working practices had become 'institutionalized' and although absenteeism had been reduced as a result of a move from the previous site where appalling working conditions existed, management still considered the workers to be unreliable, but the workers themselves had managed to retain control of the quality and pace of production.

A very different managerial approach was evident in the case of an optical company, which was involved in the production of spectacles to prescription. This type of work was traditionally highly skilled and involved work requiring high-precision workmanship. The managers were former craftsmen themselves, and the firm's policy was to buy the most modern precision machinery possible. A whole range of precision machinery had recently been computerized, thus effectively deskilling the surfacers and glazers.

Given all this deskilling, one might have expected management to exploit the control possibilities of the quality and quantity of production rather than relying on the judgement, skill and discretion of the work-force. However, management still retained a group bonus scheme and allowed the work force to establish a working liaison with the bonus department to regulate their own output. Management even persuaded the surfacers to do some of their own computer programming, and introduced job rotation within both the surfacing and glazing sections. Alongside the rotation of jobs was a policy of careful personnel selection which ensured not only a minimum level of competence but also promotion possibilities.

Source: Barry Wilkinson, *The Shopfloor Politics of New Technology*, Heinemann Educational Books, London, 1983.

As Wilkinson writes:[18]

> Comparing this firm with the plating company displays the folly in the notion that deskilling and increased management control over production processes necessarily leads to increases in efficiency

Even with deskilling management was able to shape the social and technical organization of work and ameliorate the changes which would otherwise have led to a demoralized work-force.

It is worth pointing out that the optical manufacturing case outlined above does not necessarily imply an altruistic policy on the part of management. Such policies provide employers with a valuable increase in flexibility of deployment which is likely to permit more economical manning. The effect of such policies is to create a much sounder basis for long-term employment opportunities at a time when employment prospects for those with particular skills are gradually being eroded. In this way, management is able to foster the commitment of workers to the company, and hence enhance managerial control. As we saw in chapter 1, this kind of policy bears some resemblance to that practised in many large Japanese enterprises.

Unfortunately, the management philosophy of regarding new technology as a means for downgrading and even replacing labour is all too common. Wilkinson provides another case concerning a rubber moulding company. Here, the operators became familiar with a new set of moulding machinery soon after it had been installed, and were performing wider functions than was expected of them – for example routine maintenance and process control setting. They were thus able to control the new machinery as well as their supervisors, and when a new piecework scheme was introduced they used these newly acquired skills to bargain for higher rates. Management 'learned their lesson' from this experience and subsequently introduced changes in work organization which effectively transferred the setting and adjusting functions of the moulding machinery to the process controllers in the office. It was intended that as soon as new machinery was installed in the future it would be set up in a development area well away from the shop floor.

Another example of the degrading of skills philosophy can be found in the final report of the Machine Tool Task Force, a 2½-year project funded and guided by the US Air Force.[19] This publication offers some key insights into the design criteria of new technology and contains five volumes filled with engineering detail. The assumption is made that increased productivity and profitability will flow from a work environment in which all variables have been eliminated. *Human participation* is singled out as a key variable, and one of the central conclusions of the report is that machines should be designed to 'reduce operator involvement'. One way to achieve this is to 'simplify controls to allow use of lower-skilled labor' and to 'approach unmanned operation'. The report recommended that industry should 'reduce the skill levels required to operate or maintain certain machine tools'.

Another American publication, this time a highly respected trade journal, reported in its September 1981 issue the results of a survey of managers in the metal-working industry concerning flexible manufacturing systems (FMS):

> . . . labor's role in manufacturing, particularly as regards control over production rates and product quality, is being thoroughly re-examined. Workers and their unions have too much say in the manufacturer's destiny, many metal-working executives feel that sophisticated FMS's can help wrest some of that control away from labor and put it back in the hands of management where it belongs.

We have already mentioned that new technology has the effect of eroding the traditional boundaries between occupations. Wilkinson illustrates the conflict that can arise between different occupations when CNC machines are introduced on to the shop floor. The case was that of a major machine tool manufacturer and there was conflict between planners, programmers and CNC operators.

The planners were the office staff who provided the planning sheets containing information on methods of manufacture (in the case of machining, the type and order of cuts, etc.). They were frequently referred to by the operators as 'office wallahs' whose 'lack of shop floor experience' rendered their planning instructions highly suspect and subject to correction, increasing the work load of the machinist and increasing the perceived importance of management's reliance on the machinists. The planners' work also overlapped with the programmers, and this led to frequent complaints from the programmers that the planners' 'lack of programming experience means their planning sheets are not always appropriate to the needs of CNC machining'.

With the transfer to computer control the planners' work went first to the programmer, whose program had to be written according to the specifications laid down by the planner. The planner had complete discretion in deciding which jobs were suitable for the CNC machines: increasingly, the programmers were starting to question this discretion, claiming that they themselves were far more able to understand the capabilities of modern CNC equipment. In some cases, the programmers even changed the plan according to their own specifications. Programmers were also encroaching upon the traditional role of the jig and tool draughtsmen. When special tools were needed ' . . .we design and draw tools ourselves rather than getting the jig and tool draughtsmen to draw them'. Thus the boundaries between the various office functions were the subject of continual re-negotiation on each piece of work.

There was a similar negotiation and re-negotiation of tasks between programmers and CNC operators on the shop floor. There had been no rules laid down by management for programmers and operators to follow, and in the absence of such management direction, no clear-cut divisions of task allocation had been settled. The CNC operators were all skilled machinists (the shop stewards had insisted that the most skilled machinists should have priority in operating the new CNC machines) and the programmers had served a lengthy technical apprenticeship geared to the needs of computer control.

It was discovered that the CNC operators were becoming increasingly involved in *more* than just setting and minding the machines, and many of them insisted on correcting faults in the program tape – often to take account of variations in material quality. In some cases the operators even tried embellishments that improved the tool's performance. The programmers insisted that editing was their domain, and a continual battle for control was being waged. The conflict was symbolized by a key on the control cabinet of each CNC machine, which unlocked and locked the tape editing facility. The programmers would have liked to have taken this key away, having proved their own programs, but the CNC operators had succeeded in retaining full control of the machine's control panel. One operator, who had 4 years' experience on CNC centres, had taught himself to program, starting from an engineering drawing and programming the machine using the *tape editing facility*. He knew of other machinists in other companies who had picked up this same skill. Management turned a blind eye to what was going on; in fact they frequently expressed admiration for the quality of work being produced this way.

In another machine tool company which Wilkinson studied where management made strenuous efforts to keep all aspects of machine programming away from the shop floor, workers often reassumed control through undertaking instruction in programming in their own spare time and at their own expense by gaining access to control boxes with keys they had made themselves. It was observed that the use of skills by those on the spot could avoid considerable wastage of time and material when the need for unforeseen adjustments arose.

In a report prepared for the US Office of Technology Assessment (OTA),[20] one of the four case studies in the report concerned the introduction of computerized machine tools into seven small metal-working shops. The study found that centralized control of shop operations had increased, overall skill requirements had decreased, and many skilled workers complained that their jobs had become routine and boring. As one skilled operator put it:

> You don't have to think anymore. You lose touch with what you're actually doing. . . . You've got a set of documents, you've got your tapes, and you've got a set of instructions as to how to set the machine up and run it, and I believe you yourself become a robot, because you don't get to use your mind.

While computerized machine tools may be designed to minimize operator input, there are limits to the reduction of skill and worker intervention, given the enormous variability of batch manufacturing and the current state of the technology. In the seven small shops visited in the OTA study, the day-to-day reality was filled with the unexpected: bugs in parts programs, unforeseen problems with the machines, off-standard castings, and many other variables. Under these circumstances, operator intervention is critical, even though the need for many important skills is eliminated or reduced. Ironically, some of the most satisfying experiences on the job that workers spoke of occurred when machines and systems were not operating according to plan.

As Shaiken observes:

> Rather than address the boredom syndrome by re-designing work, employers often introduce more technology. To compensate for boredom and lack of operator involvement, engineers are asked to design more and more complex machines to replace the few remaining human functions that exist. The increased complexity can lead to increased unreliability.

The new technology on the shop floor can also lead to *decreased autonomy* on the part of workers and enhanced management monitoring of work. New management information systems can go beyond simply co-ordinating production: the way data are defined, collected, and used can result in new forms of control, both subtle and direct, over what workers do. This control can diminish the autonomy that both workers and supervisors have in production. One such system had been installed in one large manufacturing company visited in the OTA study. When a worker completes a job, stops work for other reasons, or punches in or out, this information is keyed into a terminal near the work area. At a touch, a supervisor can call up detailed information about every aspect of a department's activities. Using these data, a worker's on-the-job activities can be controlled as well as monitored. An important target of this system is the practice of 'banking' of work, whereby a worker accumulates a stockpile of parts by working faster than the prescribed methods in order to even out fluctuations in incentive earnings.

In other monitoring systems reported in the OTA study the computer was linked directly to the machine tool. Thus when the machine was not operating for one reason or another, the supervisor was notified and there were severe limits on the ability of workers to control the production rates on the shop floor. Because all transactions were logged in the computer, the foreman was no longer able to turn a blind eye to on-the-spot corrections of work, and thus informal arrangements between workers and supervisors were destroyed.

One of the dangers of computerized manufacturing is that it destroys the *social relationships* between different occupational groups that form part of the reality of everyday production on the shop floor. For example, conventional design methods in manufacturing are by their very nature interactive. When a designer draws a part he or she must take that drawing down to the shop floor, explain it to the machinist, listen to the machinist's suggestions and generally interact with the shop floor environment. With CAD, however, it becomes possible to model a design within the computer itself. Thus, prototypes often do not have to be built to test various design alternatives. The design can even be converted into instructions to guide a machine tool (CAD/CAM), and these instructions can be electronically transferred to the shop floor.

Despite the bleak picture which has so far been painted about the potential of new microelectronic applications in manufacturing to degrade and deskill factory workers, there is little evidence that there is any *unilinear* tendency towards deskilling stemming from the inherent nature of the new technology.[21] Jones, for example, scrutinized one sector the British engineering industry to examine the adoption of NC machines. He showed that there was considerable variation between firms, and he claimed that this confirms the importance of product and labour markets, organizational structures and trade union influence as independent influences on the forms of skill deployment.

Similarly, in the comparative study of microelectronics and manpower in manufacturing in Britain and West Germany by Nicholas, Warner, Sorge and Hartman which has already been referred to, there was little evidence of CNC systems causing deskilling amongst machine operators; in fact in many of the smaller German companies there was evidence of a *renewed interest* to train and employ skilled workers. Polarization of skills was strongly linked to batch size and size of plant, with more skill polarization evident with large batches and in larger plants. Cultural variations between the two countries also influenced the mode of technological application.

Why do firms introduce the new microelectronics technology? Do they do so primarily to increase their control over labour? It is difficult to generalize about these questions. While there are plenty of instances cited of management seeking to use the new technology to deskill and weaken workers' influence over the production process (see the OTA cases already mentioned), evidence from Rose and Jones[22] from six British firms which they investigated led them to the conclusion the concept of a 'conscious management strategy of control' may be unhelpful, if not misleading, as a guide to the kinds of policy and change affecting work organization in British work places.

Bessant[23] reported that there was rarely any defined motive on the part of management in the firms he studied. Rather, a range of several factors were used to justify new investment in new technology, with labour considerations no more important than any other factor. In a similar vein, Buchanan[24] concludes that technical change merely acts as a *trigger* to processes of management decision-making, and he emphazises that in every case he studied it was possible to identify a 'promoter' or 'champion' in the company who was instrumental in persuading other managers of the value of investment in new technology.

Conclusions

The current developments in manufacturing technology herald dramatic changes in employment levels, skills, and job design which are already taking place in many manufacturing enterprises. One of the most depressing conclusions is that the Taylorist philosophy is in many cases still being carried over to the era of microelectronics systems in manufacturing.

The use of computer-based automation that trivializes work is both humanly tragic and technically unnecessary. An alternative direction that fully uses the extraordinary abilities that people possess requires a new vision of what technology can accomplish. This vision would not counterpose computers to people; it would not relegate human abilities to being taken over by an electronically controlled system. Instead, computers would amplify rather than eliminate the unique qualities that only humans can bring to production. The design of machines reflects *social values* as well as technical needs.

The alternative is a much darker vision: the creation of a pyramid of skills that concentrates a few creative and meaningful occupations at the top, while the rest wind up with fewer skills and subject to new forms of monitoring and electronic control.

4

The Unemployment Threat

Concern about unemployment as a result of automation stretches back over a long period. Long before the microprocessor had been invented in the early 1960s a debate was already in progress about the effects of computers on employment levels. Then, as now, there was a great deal of disagreement among the experts, whose views were divided and ranged from the optimistic to the pessimistic. Despite the dissensus that exists today, we do know that there is nevertheless some degree of convergence of opinion about the relationship between automation and unemployment.

First, the experts appear to agree that the impact of automation on employment will vary considerably depending on the underlying growth rate in the economy. Those countries with higher growth rates, such as Japan, will be less likely to experience unemployment as a result of technological developments than countries with lower growth rates such as Britain.

Second, we know that any unemployment induced by new technology will be related to the rate of growth of the labour force. The level of unemployment is related to the numbers entering and leaving the labour market from time to time, and those countries which have a growing number of people seeking jobs (for example school-leavers, women hoping to return to full or part-time employment) are much more likely to experience unemployment than countries with more stable labour forces.

Third, as politicians and industrialists are quick to point out, new technological developments inevitably create transitional forms of unemployment, but over the longer term new jobs will be created in new industries and processes. These new jobs may require different skills and knowledge, and problems are likely where jobs are destroyed faster than the new ones are created – leading to a mis-match of skills, and the possibility of regional problems of unemployment arising as a result of the decline of older industries in particular areas. Thus there will be a need for re-training and re-education in order to encourage mobility of workers to new job opportunities in other areas – either by relying on market forces or by the government itself assuming the responsibility for managing the labour market accordingly.

Which Jobs are most at Risk?

There have been very few studies so far which have sought to identify particular occupations which are most at risk as a result of the introduction of the new microelectronic-based technology. We have to treat the results of such studies with some caution because the occupations concerned will not necessarily disappear as such but might simply be changed in some way – either by a form of deskilling or by being subsumed into other tasks which were formerly carried out by other occupational categories. It is also important to differentiate between the short-term and long-term changes involved.

Given below are two tables which list a number of occupational categories based on ILO specifications; table 4.1 deals with occupations which are changed considerably in the short term and table 4.2 the longer term. It should be stressed that these tables provide no more than an over simplified view of what is in reality a complex picture.

Table 4.1 Occupations changed in the short term

Occupation	*Cause of change*
Printers and typesetters	Integrated text processing, control of presses
Welders	Robots and handling equipment
Electrical fitters, etc.	Automatic assembly units, more company integration
Technical draughtsmen	CAD/CAM
Laboratory technicians	Automatic analysers
Data processing workers	Software technology developments
Technicians	CAD/CAM
Engineers	CAD/CAM
Secretaries, typists etc.	Text processing, automated office
Metalworkers	CNC machines, robots

Table 4.2 Occupations changed in the long term

Occupation	*Cause of change*
Office managers and assistants, sales staff, administration employees	Automated administration with linked decentralized systems
Sales staff, wholesale and retail	Computerized cash registers, product code, integrated storage and stock control
Telephone and telex staff	Automated exchanges
Textile occupations	Automatic machinery
Warehouse workers	Automated warehouses
Bank and insurance experts	Automated payments, electronic funds transfer
Accountants, cashiers	Automated payments, electronic funds transfer
Dispatch workers, checkers	Automated packaging and checking
Postal staff	Electronic mail, etc.
Supervisors and foremen	CAM and CNC machines
Toolmakers	Programmed machine tools
Electrical mechanics	Greater integration of components
Mineworkers	Computer-controlled machinery.

These tables by no means constitute an exhaustive list of all the occupations that are likely to be affected by the new technology. In fact, as telecommunications becomes increasingly linked to computerized systems and devices these occupations (and many more) will be increasingly affected.

New Technology, Growth and Unemployment

In the aftermath of the second world war, successive governments in all the Western economies pursued policies of full employment, and there was a widespread determination never again to return to the deprivations and depression of the 1930s. For about 25 years after the end of the war all Western countries experienced a period of full employment, growth and prosperity. Those who had warned about the dangers of automation destroying jobs during the 1950s and 1960s had been proved wrong by the course of events. Far from destroying jobs, new technological innovations led to a demand for products and services and were associated with the growth of new job opportunities. In addition, in many Western economies there was a growth in employment opportunities stemming from the expansion of the public sector and the increased role of government in providing services to the public. The long-term tendency in all industrial societies – a trend which accelerated during the post-war period – has been for the majority of the working population to be employed in the tertiary sector of the economy, chiefly at the expense of manufacturing and agriculture (see table 4.3). Indeed, the service sector has constituted the majority of the employed population in Britain since the 1960s, and in the United States well before that date.

High rates of economic growth throughout the West slowed down dramatically during the early part of the 1970s and there has been a widespread increase in unemployment in all Western economies since 1975. The generally accepted opinion is that high levels of unemployment are likely to persist and might even worsen in the foreseeable future. Although this rise in unemployment started well before the widespread application of the new microelectronic-based technology, there is a real danger that the spectacular advances in information technology – particularly arising from the marriage of ADP and telecommunications – will lead to even higher levels of unemployment.

Does this all mean that the 'triangular' relationship between employment, economic growth and technological development – whereby although technological innovation caused job loss there were nevertheless alternative job opportunities available as a result of the growth in productivity and the creation of new wealth – has now been replaced by a simple causal relationship between technology and employment levels? Can we therefore expect to see unemployment continuing to grow as the new technologies are gradually applied throughout industry?

Unfortunately, it is extremely difficult to give a definite answer to these questions. In the first place, the structural changes in the pattern of employment which many now refer to as 'de-industrialization' – i.e. the transfer of employment from producing industries, particularly manufacturing, to the service industries – is considerably oversimplified and disguises the fact that the decline in employment has by no means been uniform across all producing industries; nor has employment increased uniformly across service industries. Employment in certain British manufacturing industries has been hit harder than in others; for example in steel, cotton, footwear, motorcycles, shipbuilding and clothing. Employment in certain service industries has also declined; for example in transport and distribution. In these cases, the decline in employment

Table 4.3 Structure of civilian employment: women and wage and salary earners (in percentages)

	Agriculture			*Industry*			*Services*			*Women*			*Wage and salary earners*		
	1960	1973	1981	1960	1973	1981	1960	1973	1981	1960	1973	1981	1960	1973	1980
Australia	10.3[c]	7.4	6.5	39.9[c]	35.5	30.6	49.8[c]	57.1	62.8	—	33.6	36.3	83.5[c]	86.4	84.1[d]
Austria	24.6	16.2	10.3	40.3	40.6	40.1	35.1	43.2	50.0	—	38.4	38.2	—	75.4	82.9
Belgium	8.7	3.8	3.0[d]	46.8	41.5	34.8[d]	44.6	54.7	62.3[d]	30.7	34.0	35.9[d]	73.8	82.9	83.4
Canada	13.3	6.5	5.5	33.2	30.6	28.3	53.5	62.8	66.2	26.8	35.2	39.7	81.2	90.1	90.1
Denmark	18.2	9.5	8.3[e]	36.9	33.8	30.0[e]	44.8	56.7	61.7[e]	31.8	41.1	43.6[e]	76.4	81.5	83.9[e]
Finland	36.4	17.1	11.1	31.9	35.7	34.8	31.7	47.1	54.1	44.8	46.1	47.6	63.7	80.8	86.3
France	22.4	11.4	8.6	37.8	39.7	35.2	39.8	48.9	56.2	35.2[f]	36.0	38.0[d]	69.5	80.7	82.9
FRG	14.0	7.5	5.9	48.8	47.5	44.1	37.3	45.0	49.9	37.8	37.2	38.7	77.2	84.2	86.2
Greece	53.8[c]	38.9[h]	30.8[e]	18.5[g]	26.3[h]	30.0[e]	27.7[g]	34.8[h]	39.2[e]	32.3[g]	27.5[h]	29.7[e]	33.5[g]	—	48.6[c]
Ireland	37.3	24.8	19.2[d]	23.7	30.9	32.4[d]	39.0	44.2	48.4[d]	26.6[g]	26.6[h]	28.5[d]	61.4[g]	70.6	74.3
Italy	32.8	18.3	13.4	36.9	39.2	37.5	30.2	42.5	49.2	30.1	28.7	32.3	58.4	69.4	71.5
Japan	30.2	13.4	10.0	28.5	37.2	35.3	41.3	49.3	54.7	40.7	38.5	38.7	53.4	68.7	71.7
Netherlands[a]	11.5	6.8	6.0[d]	40.4	36.2	31.9[d]	48.2	57.0	62.1[d]	—	—	—	78.1	84.1	86.8
Norway	21.6	11.4	8.5	35.6	33.9	29.8	42.9	54.7	61.7	29.0	36.6	41.4	74.3	83.1	86.4
Portugal	42.8	34.8[i]	28.5[d]	29.5	34.5[i]	36.0[d]	27.7	30.7[i]	35.5[d]	18.7	40.0[i]	38.8[i]	74.3	65.9[i]	66.7
Spain	42.3	24.3	18.2	32.0	36.7	35.2	25.7	39.0	46.6	—	28.0	28.6	61.0	67.2	69.6
Sweden	13.1[j]	7.1	5.6	42.0[j]	36.8	31.3	45.0[j]	56.0	63.1	36.1[j]	40.8	45.9	86.9[k]	90.8	92.0
Switzerland	13.2	7.7	7.0	48.4	44.1	39.3	38.4	48.1	53.6	—	34.0	35.2	85.5	—	—
Turkey	81.1	64.5	60.4[d]	8.6	15.1	16.3[d]	10.2	20.4	23.3[d]	45.2	—	—	13.6	29.7	34.2[d]
United Kingdom	4.1	2.9	2.8	48.8	42.6	36.3	47.0	54.5	60.9	34.4	37.6	40.3	92.7	92.1	89.8
United States	8.3	4.2	3.5	33.6	33.2	30.1	58.1	62.6	66.4	33.3	38.5	42.8	83.9	90.3	90.6
OECD[b]	21.7	12.1	10.0[b]	35.3	36.4	33.7[b]	43.0	51.5	56.3[b]	34.3	36.2	38.6[b]	70.5	80.5	82.0

Note: Based on ILO definitions and reproduced in *World Labour Report, 1984*, ILO, Geneva, 1984.
— = not available [a] – in work years; [b] – estimated; [c] – 1964; [d] – 1980; [e] – 1979; [f] – 1968; [g] – 1961; [h] – 1971; [i] – 1974; [j] – 1962; [k] – 1967.

was as much to do with a reduction in demand and competition as it was to do with technological advance.

Second, it is equally difficult to isolate the employment displacement effects arising from technological innovation from other factors. The economic context, internal and external competition, raw material markets, factor prices, changes in demographic aspects, etc., will all play a part.

Economists have been trying for years to develop models (which involved drastic simplification of real-life complexities) to explain the complex interaction between technology and employment levels. The large number of variables involved makes any detailed forecasts almost impossible. Any such forecast would require very detailed analysis of every economic sector, and the studies that have been carried out so far have produced confusing and often contradictory results. Such forecasts either did not consider all the relevant factors involved, were based on questionable assumptions, or simply took the effects of one situation, one product, one branch, one country, and transferred that situation to a different case.

The drastic decline in employment in the Swiss watchmaking industry is frequently quoted as an example of the effects of microelectronics on employment. Here, microelectronics caused a reduction of around 46,000 employees during the mid-1970s; in West Germany the number of employees in watchmaking was reduced by 40 per cent. There are numerous examples of other industries which can be cited as examples of employment levels being reduced as a direct or indirect result of microelectronics. However, many of these examples refer to products which contain a relatively high proportion of information and control components in comparison to the average proportion in other products.

Moreover, one must also recognize that such employment reductions do not give any indication of the extent of increased demand for new products as a result of improved quality and lower cost. There will also be a need for more people to be employed in making components for the new products, quite apart from the multiplication effects of jobs being created elsewhere as a consequence of the opening up of new markets.

Third, a great deal may depend on the *rate of diffusion* (that is, the spread) of the new technology. All major new technologies throughout history, such as electricity or the steam engine, have required a gestation period before their full social and economic impact was reached. In the early period of a new technology, even where the potential technological capability is well established, there will inevitably be design problems, high costs, performance unreliability and a marked shortage of well-trained personnel who are familiar with the new technology. Not only will these factors inhibit the diffusion rate, but also management itself will be suspicious of a new innovation until there is sufficient evidence that it is likely to be of benefit to them. It takes a considerable period of time before society and industry are affected by a far-reaching major technology.

In this context, it can be argued that the diffusion process of microelectronics is likely to be much shorter than with previous major technologies. Chips are small, they are more reliable and they are becoming more powerful and cheaper at an extraordinary rate. There is already a large number of people who are skilled in electronics and systems analysis. Whilst we are still in the embryonic stage of the 'information revolution', there is sufficient evidence of the severe job-displacement effects in a whole range of industries to warrant grave concern. Examples can be cited of areas such as car manufacturing, telephone operation, watchmaking, printing and publishing, tax collection, clerical work, electronic assembly, typewriter manufacture and in relation to the check in employment growth in some service industries such as banking and insurance.

It is very difficult to predict the diffusion rate of information-related technology. The question is as much bound up with economics as it is with the technology itself, in the sense that a great deal depends on the economic capability to take advantage of the opportunities offered by the new technology. Who can predict with reasonable accuracy what the economic situation will be in the years to come?

It can be argued that the diffusion rate of microelectronics is likely to be high because it can enable the automation of work to be extended to a vast range of new areas. In the past – particularly during the early part of the twentieth century – automation of manufacturing was restricted to mass production industries. As we saw earlier, the advent of microelectronics enables automation to be extended to small-batch manufacturing (which accounts for over 70 per cent of worldwide manufacturing), through the use of CAD and CAM. Similarly, automation in offices has been, until very recently, restricted to typewriters, duplicating machines and photocopiers. Now many experts discuss the inevitability of the 'paperless' or 'electronic office'.

So much depends on the future directions of world trade and the way industrial structure is altered. The more revolutionary the technology, the greater will be the impact on products, markets, distribution and individual enterprises. Yet even if the new technology (does) create more jobs in the long run than it destroys, it does not necessarily follow that full employment will ever return to Western industrialized societies. As companies diversify and transfer their activities across international boundaries, the new technology could create more jobs in underdeveloped countries than it does in the advanced Western economies as a result of a radical shift in the international division of labour. The new information technology can serve to raise the stakes in international trade competition and the costs of losing out are potentially very high.

The history of technological development shows that the changes which are brought about by radical innovations do lead to the creation of completely new industries and occupations; conversely, they can lead to the displacement of labour through increases in efficiency in existing processes of producing and delivering goods and services, assuming that the demand for these goods and services does not increase sufficiently. It is often assumed that these processes are automatically in balance, but it is misleading, as we will see later in this chapter, to oversimplify the process whereby this is achieved.

Some Forecasts of Likely Employment Effects

The accuracy of the various forecasts that have been carried out depends not only on the diffusion of the new technology and the way in which it is applied, but also on the competitive position of the economy as a whole, the level of aggregate demand available to a country's industries and the balance between employment in manufacturing and services. Many of these points are dealt with in more detail in the rest of this chapter. One study from the Federal Republic of Germany (see Box 4.1) carried out by the Institute of Labour Market and Occupational Research analysed technical change in six sectors of activity: synthetic material, wood, food processing, metalworking, printing and the retail trade.[1]

At first sight, the table in Box 4.1 appears to suggest that the employment effects of technical change are positive, but in fact this is mainly accounted for by the expansion of activities. On closer examination, by looking at the innovation and rationalization variables alone, it is clear that technological change has a fairly direct negative effect on the sectors considered. Moreover, the surveys ended in 1977 – just at the time when we would expect that the accelerated impact of information technology, combined with the economic recession, would have a profound negative impact, especially in sectors such as printing.

BOX 4.1

In 1,600 representative enterprises the Institute analysed about 3,000 technical changes subdivided into four categories. The first category was one of *innovation* – a phase in which fundamental research results in new products or processes. This phase includes new product installations and techniques, the use of new materials and the application of new energy sources and electronic data processing. The second phase was one of *rationalization* – making it possible to produce conventional products more efficiently while the use of production factors remains equal or diminishes (e.g. the steel industry or the car industry). The rationalization phase involves mechanization, automation, organizational changes and transfer of production activities. The third category involved *expansion of activities,* while the fourth category was one of *reduction of activities.* Each of the expansion and reduction types involve either a reduction or expansion of activity without a change in technology or products. These phases depend on the general level of economic activity and its position in the economic cycle.

The results of the German study are set out in the table below.

Employment effects of the four types of economic change in the FRG (as a percentage of the 5 – 6 million employed in the undertakings analysed

Year	*Source*	*Innovation*	*Rationalization*	*Expansion*	*Reduction*	*Total*
1970	*Synthetic material*	*+1.01*	*−0.05*	*+2.07*	*−0.20*	*+2.83*
1971	*Wood*	*+0.81*	*+0.09*	*+0.54*	*−0.21*	*+1.23*
1972	*Food processing*	*−0.06*	*−0.74*	*+0.92*	*−0.31*	*−0.19*
1973	*Metalworking*	*+0.43*	*−0.09*	*+1.40*	*−0.15*	*+1.59*
1975	*Printing*	*−0.38*	*−0.37*	*+0.24*	*−0.39*	*−0.90*
1977	*Retail trade*	*+0.21*	*−0.37*	*+2.16*	*−0.34*	*+1.70*
Average		*+0.34*	*−0.26*	*+1.22*	*−0.27*	*+1.03*

Source: W. Dostal: Bildung und Beschäftigung im technischen Wandel (Beiträge zur Arbeitsmarkt und Berufsforschung No. 65) Nuremberg, Institut für Arbeitsmarkt und Berufsforschung der Bundesanstalt für Arbeit, 1982.

In a paper which was prepared for the OECD (see Box 4.2) for an Inter-Governmental Conference on Employment Growth in the Context of Structural Change (February 1984), a forecast was produced for manufacturing employment for the major OECD countries (Canada, United States, Japan, France, West Germany and the United Kingdom).

BOX 4.2

Two different scenarios were considered in the OECD study. A relatively optimistic one which assumed that both output and investment continued to grow at the same rate as over the post-war period (1955–81), and a more pessimistic scenario which assumed that output and investment growth rates were similar to the 1970s (a period of much lower growth rates in nearly all the Western economies). The *optimistic* scenario produced forecasts on employment gains/losses compared to 1982 levels in millions by 1990 and 1995 respectively:

Canada +0.13 and +0.30;
United States −0.31 and +0.47;
Japan +3.35 and +5.69;
France +0.51 and +0.85;
Germany −0.01 and +0.85;
United Kingdom −0.48 and −0.86.

However, the *pessimistic* scenario produced the following figures:

Canada −0.01 and +0.09;
United States −1.56 and −1.70;
Japan +0.37 and +0.99;
France −0.18 and −0.24;
Germany −1.04 and −1.38;
United Kingdom −1.28 and −1.97.

There are a number of points that ought to be stressed in connection with these forecasts. In the first place they deal with manufacturing employment only, and many job losses will occur in the offices and in the service sector.

Second, future employment prospects in the OECD countries will depend on four main factors, even assuming reasonably stable, if sluggish, economic growth. These factors are the extent to which the restructuring of economies towards more advanced technology-based industries can be achieved; the extent to which employment opportunities will develop in the service sector; the extent to which education and training provisions match up to job opportunities; and finally, factors which affect the demand for jobs in the labour market – that is, the number of people needing or desiring employment, and the various measures which are available to governments to deal with this demand, such as raising the school leaving age or lowering the retirement age.

In addition, as we have seen already, forecasting of technologically induced unemployment is plagued by very real difficulties in being able to isolate technology from a whole range of other factors. In France, a special study undertaken in 1978 for the French President concluded that there would be a 30 per cent displacement of office workers in banking and insurance companies and a 'massive' impact on office work. This has significant implications for women workers, who tend to be concentrated in those sectors.

Perhaps the most well-known forecast of employment levels in Britain as a result of the impact of microelectronics is that of Clive Jenkins and Barrie Sherman (1979).[2] They spoke of a job reduction of 5 million in the United Kingdom over 25 years, which was 23 per cent of the 1978 labour force. This prediction was based on the assumptions that the manufacturing sector would continue to decline and that maximum use would be made of the new technology. While many derided this forecast as being wildly pessimistic, the dramatic rise in unemployment since 1979 has shown that Jenkins and Sherman *under-estimated* the 'de-industrialization' of the British economy under the Thatcher administration.

Another British study by Barron and Curnow (1979)[4] estimated that there would be a continuing level of unemployment for the 1980s of from 10 to 15 per cent if the application of microelectronics continued to take place in a low-growth economy. One group of workers who were seen to be particularly vulnerable were women workers who were located in office jobs. An Equal Opportunities Commission Report published in 1980[5] examined the impact of new office technology on female employment and concluded that by 1985 some 21,000 typing and secretarial jobs would have been displaced by new technology, with a peak of displacement being reached by 1990 when some 17 per cent of such jobs were expected to have been lost. Unskilled and semi-skilled clerical workers would be at an even greater risk than secretaries and typists. Those women

BOX 4.3

In a much-quoted survey which was carried out by researchers from the University of Michigan and the Society of Manufacturing Engineers,[3] it was predicted that by 1990 the major centres of production in the USA would run on a 32-hour week (or four 8-hour days). Changes in the nature, duration and allocation of work would be negotiated with the work-force in the following stages:

* from 1980 a new hierarchy of skills, ever more centred on the creation, realization and maintenance of automated equipment, would come into being;
* by 1985 20 per cent of the workers currently employed in assembly would be replaced by automated systems;
* by 1987 20 per cent of industrial jobs would be redesigned and 15 per cent of assembly systems automated;
* by 1988 50 per cent of the labour force employed in assembling would have been replaced.

The process was due to accelerate after that. According to a study presented by the Stanford Research Institute to the United Auto Workers (UAW) in March 1979, 80 per cent of manual work would be automated by the year 2000 which, at present working hours, would amount to the elimination of 20 million manual jobs from the present total of 25 million.

who were engaged in light repetitive assembly, where the potential for automation is likely to be high, were also likely to be displaced in a major way. Women's jobs in banking were also vulnerable.

In Europe, the European Trade Union Institute (1980)[6] predicted substantial losses of employment due to microelectronics in the manufacturing, finance, communications, and transport sectors. The most drastic effects in manufacturing were expected to be within labour-intensive assembly line production where new factory layouts, coupled with computer-controlled operations, warehousing, and inspection, could replace many jobs which at present are performed by semi-skilled, skilled and supervisory grades. Examples of such cases included car manufacture, the production of fridges, washing machines, cookers, and the manufacture of cash registers.

New Jobs in the Service Sector?

The service sector of the economy is a key sector for any hopes that we may have about providing new job opportunities to replace those which are undoubtedly going to disappear in the years to come. Many experts still assume that even though jobs will be increasingly lost in manufacturing and other traditional industries, the expansion of the service sector would easily maintain employment levels. Many still believe that the current levels of unemployment are just temporary products of the worldwide recession. The coming of the 'post-industrial society' is seen as providing new forms of job creation.

However, there is no guarantee that any such smooth transition will take place. The continued expansion of the public sector and government employment is at best uncertain and the new information technology will displace jobs in the service sector as well as in traditional industries. Before considering what the likely possibilities are in this regard, it is worth looking at the way that the service sector has expanded in recent years. Employment in the service sector in the ten countries of the European Community grew from around 38 per cent in 1958 to around 56 per cent by 1982. In the United States the figures were 57 per cent and 68 per cent over the same period, whilst in Japan service employment increased from just below 40 per cent to around 56 per cent.

A study of the prospects for employment in the 'post-industrial society' has already been undertaken by Jonathan Gershuny and Ian Miles, both of the Science Policy Research Unit at the University of Sussex in Great Britain.[7] They argued that the new telecommunications, computing and information storage technologies present the technical inputs for a new wave of social innovations in entertainment, information, education and even medical services – and these social innovations could produce economic effects at least as substantial as those experienced in the 1950s and 1960s.

According to Gershuny,[8] goods and services in the economy are both ultimately consumed as services, and he calls such ultimate services 'the final service' function. To illustrate his argument, he cites the purchase of a rail ticket from destination A to destination B as the consumption of a service, which is intangible, whereas the purchase of a car represents the purchase of a tangible good. The latter purchase enables the user to travel between A and B (and other destinations) whenever he or she wishes, and can be seen as indicative of a move towards the 'self-service' economy. Similarly, consumers can either pay for the provision of laundry services, or buy and use a washing machine; go to the theatre, or buy and use a television or video recorder.

Gershuny argues that the last 25 years has seen a rather regular pattern of change in the mode of provision of some final service functions. As he puts it:

> In the domestic, entertainment and transport spheres, most households have shifted from a predominantly serviced to a predominantly self-serviced mode of provision. This change can be explained in part by technological developments (in the design and production of small electric motors, semiconductors, the internal combustion engine etc.), in part by the development of particular sorts of material infrastructure (electricity generating networks, road and telephone systems), and in part by economic pressures [particularly the rising relative cost of purchasing final services] which reflect the technological changes. And we can also foresee new sorts of changes in the mode of provision of services, particularly in service functions which have not in the past been affected in this way, as a result of new technologies, and new infrastructural provisions.

Gershuny's argument is that the 1950s and 1960s saw a significant decline in jobs in the 'final services' sector (with the notable exceptions of education and medicine). There was, for example, substantial job loss in transport provisions, out-of-home entertainment and laundries – all examples of industries which were directly hit by the shift towards 'self-service' consumption patterns. True, there was also job generation in manufacturing industries to offset this job loss in much of the 'final services' sector (for example an increased demand for washing machines, refrigerators, motor cars and televisions); there was also a need to provide an infrastructure to support such 'self-service' consumption patterns (for example electricity systems, road networks and telephone links).

Hence, the new markets for the products associated with consumer durables, motor manufacturing, and consumer electronics fuelled the growth in the materials manufacturing and capital goods sectors, and this provided the social surplus that could be devoted to the educational and medical services. As Gershuny and Miles note:[9]

> The self-servicing trend can be seen as the motor of economic growth, establishing or extending markets for consumer goods, and hence, through the Keynesian multiplier, extending demand, and accordingly employment, throughout the economy. It may be that this consequence, on employment via the level of effective demand, is more important than either the direct job displacing or the job generating consequences of self-servicing.

The trend towards self-servicing during the 1950s and 1960s which has just been described has continued, and is now connected with a new wave of social innovation which is based on the technological developments in microelectronics and telecommunications. The job prospects in the future 'information economy' will thus depend on the arithmetic of jobs generated and lost by the process of substitution between final service workers and intermediate service workers–'software' workers, in the broadest sense of the term. Not only will new 'software' or 'information' industries be created but these industries will have to be supported by a new infrastructure (in the form of telecommunication networks, cabling, etc.). What will this mean for job prospects in various sectors in the economy?

Jobs in the Office Sector

Office employment has been a large growth sector throughout the post-war years. The OECD countries experienced a growth in office employment of 45 per cent during the 1970s, as against 6 per cent for workers in general. There are also indications that APT (administrative, professional and technical) jobs increased at a rate more than twice that of other groups, including clericals, in France, Italy and the UK.

Because of statistical variations in classification between countries it is difficult to give accurate figures of the proportion of total employment accounted for by APT and clerical workers, but one estimate suggests that it was around 40 per cent in Britain and France by the late 1970s, and around 30 per cent in Ireland and Italy. Moreover, the ratio of APT to clerical workers in the OECD countries has increased throughout the 1970s and there are few women occupying positions in APT posts.

It is very difficult to measure productivity accurately in the office sector but there is general agreement that, as compared to manual workers, the productivity increases of office workers have been very small during recent years. In addition, capital investment per employee is much lower among office workers than it is for their manual counterparts (one estimate puts this at a ratio of 1 : 12). Thus the office sector is ripe for technological innovation, although it may be the increased need for operational efficiency which information technology will bring rather than considerations of labour cost savings *per se* that will provide the motor force for moves towards office automation.

It is much more accurate to conceive of the office as an *information-processing system* in order to plot the likely employment effects of the new technology. Information goes into the system which, together with ready-stored information, is transformed by human labour and office equipment into 'rule-governed' output information. The rules are formulated and revised by management and may even be (more or less) embodied in the physical technologies which are used. Information inputs may be of a routine nature (for example stocktaking) or motivated by less predictable factors (for example consumer enquiries); the information inputs and outputs may be in the form of telephone calls, letters or invoices, and may relate to customers, suppliers, various authorities or to the main production, storage and distribution activities of the industry in question.

New technology can be applied in different ways to the various interfaces of this information-processing system. For example, data production and analysis (accounting and monitoring of performance), and reproducing data in a form which is suitable for output, will be affected by such transformations, as well as by innovations which are specific to these activities themselves. Most of the innovations which are likely to transform office work were outlined in chapter 2, but the possibility of viewing office work as a sector for future employment creation (or job loss) can be discussed here.

A great deal depends not only on the diffusion of the new technology but also on the future demand for office services. Any such surge in demand for office services in the coming decades must be considered in relation to the potential improvements in productivity which will result from the new technology. Moreover, all the indications are that there will be *less* demand for office services in the future because the increasing automation in manufacturing will not require human co-ordination in offices between different steps in the production process that are currently involved. Instead, the technology which is now available will be able to co-ordinate the manufacture, design and control functions.

The demand for office services will vary from one industry to another, as will the rate of diffusion. Much will depend on the *intensity of the information flows and activities* that take place in particular offices. Possibly, there will be a trend towards more *concentration* in the provision of office services, as small firms find it more convenient to contract-out some of their activities to larger centres of information-processing.

Office Jobs Outside the Service Sector

How are 'service' workers in the primary and secondary sectors likely to be affected? Here, there will probably be a greater impact on employment levels arising from the effects of information-based technology on the *manufacturing* function by the use of CAD, CAM and FMS than there will be from the use of new office technology as such. In other words, clerical workers in manufacturing plants are less likely to be affected (at least in the short term) than their manual colleagues. In any case, office sites in manufacturing plants are much smaller than head offices and therefore offer less scope for much division of labour.

However, the displacement of manual workers in the same enterprises by CAD, etc. will not necessarily result in equivalent job creation of technical posts. In head offices of these companies, of course, there is bound to be extensive automation and job loss.

Producer Services, Sales and Distribution

In producer services (those firms which sell intermediate services such as consultancy, and legal advice to other producers etc.) there is ample scope for adopting new technology for routinized work, and the size of the enterprise is much less important.

Another area of employment in the service sector is what is known as sales and distribution occupations. These include travelling sales representatives, sales office staff, sales engineers, stock controllers and warehouse personnel. These occupations will be affected not so much by the technology, but by the changes that are likely to take place in the marketing and distribution of the products with which they are associated. Microelectronic equipment requires much less servicing and maintenance and can be updated by modular replacement from time to time. In addition, improvements in communications and organizational changes will enable significant enhancements in scheduling of the activities of sales personnel by 'briefcase computers', etc.– thus enabling a much greater degree of supervision to be exercised over their jobs. Although automated warehousing is very expensive, it can be expected that before too long substantial reductions in staffing will take place because it can be easily linked to automated stock control, sorting and computerized transport planning.

The Information Technology Sector

There is, however, one area where we can expect employment to *grow* as a result of information technology. Those services which are directly associated with information technology itself are expected to expand considerably, as the demand increases for 'software' workers, such as programmers, systems analysts, word processor operatives, information managers and computer consultants. Over the next two decades this sector of employment will undoubtedly be a major growth area, but hopes for future job creation in this sector must be tempered by two reservations.

First, when the so-called fifth-generation computers emerge (see Box 4.4), the scope for employment growth will be considerably reduced. Secondly, not all new jobs created will necessarily be in the advanced industrialized countries – many will be in the rapidly growing countries of the Third World, such as Taiwan and South Korea.

Banking, Insurance and Professional Services

The banking and insurance sector has grown significantly in the post-war years, but recently employment growth has been checked. Information technology is particularly suitable for application in this sector because the rapid transmission of data, much of it in numeric form, enables dramatic improvements in services and productivity. Innovations such as point-of-sale terminals and electronic fund transfer will lead to reductions in clerical

BOX 4.4

Fifth-Generation Computers

Fifth-generation computers are now being developed by the Japanese. The Fifth-Generation project identifies three main priorities: first, computers should become easier to use by everybody; second, software development should become more reliable and efficient; finally, hardware must give better performance and a wider range of functions. Essentially, such systems will be new forms of artificial intelligence, or intelligent knowledge-based systems, which will contain the distilled wisdom of people. The software of fifth-generation computers will provide for knowledge bases to act in similar ways to human logic and reasoning, deducing or inferring on the basis of previous information and experiences; that is 'thinking' for themselves.

and routine administrative functions, as well as a reduction in sales agents. There is little doubt that the future for the finance sector in job terms is one of 'jobless growth'.

Similar employment expectations apply to what can be called professional services. These occupations include research and development, advertising, marketing, data processing, accountancy, legal services, consultancy and architecture. Many routine aspects of these occupations can readily be automated (and deskilled) by computerized information searches, design software and software packages. On the other hand, there may be a growth in the activities associated with such organizations as consumers increasingly make use of them for advice on personal computer packages, education, travel, garden layout and house decor.

Communications and Telecommunications

One of the features of the post-war era has been the rise in personal mobility, in which new road networks have been constructed to cater for the rise in car ownership. An interesting possibility of the information society is the prospect that the need for transport will decline as people no longer need to be employed in enterprises as such and can carry on their work at home. Tele-conferencing enables business communications to be carried out without the need for executive travel. Much of the discussion about such a prospect has been over-dramatized and the need for transport will continue to be substantial.

However, new technology does enable significant innovations to take place in this area as a result of such things as improved communications equipment, better scheduling of routes, automated bookings via telecommunication networks and automatic ticket control. The demand for letter post will continue to be buoyant, at least in the short term, and it will be some time before electronic mail replaces the need for postal services. The future of transport is clouded by questions of energy, government transport policy and the nature of business organization several decades from now. There is likely to be a continued decline in employment in the transport sector as the infrastructure of society continues to change.

One important sector that has not yet been dealt with in any detail is the communications sector itself. This sector lies at the heart of the information revolution and comprises the physical transportation of information (mail) and parcels, telecommunications and the electronic transmission of information (sometimes reproduced in hard copy). Gershuny and Miles point out that this sector accounts for 2% of total EEC employment, with about two-thirds in postal services and the remainder largely comprising telephone services. In a study carried out in 1980 for the Union of Communication Workers in Britain by Walsh et al.,[10] it was shown that a large proportion of mail services involved business communications; 41% of British letters concern finance, while another 10% involve advertising and periodicals; only 16% of letters are between private individuals, while 36% are business to business. Considerable progress has been made in automating postal sorting, and several studies show that there will be substantial job loss as a result of increased automation. There are possibilities that some of this job loss might be offset by a commensurate growth in parcels traffic as 'tele-shopping' and telephone ordering increases.

The telecommunications sector is often assumed to be a growth sector in employment terms. In fact, overall employment in telecommunications has declined throughout the 1970s both in the UK and West Germany and the overall growth in EEC countries conceals substantial reductions in the numbers of telephone operators, with an increasing proportion of engineering posts. One German study[11] suggested that there would be a loss of around 20,000 mail handlers during the 1980s (with some job creation in telecommunications but on a much smaller scale), while in Britain British Telecom were reported by the *Financial Times* in 1984 to anticipate a reduction of nearly 50,000 of their work-force during the late 1980s. These employment predictions in the communications sector exist alongside the fact that rapid technological innovation is taking place all the time. Only a few examples are necessary to illustrate this point; teleconferencing, radio-paging, video-telephones, remote access to library services, electronic telephone exchanges, and the use of optical fibre and microwave transmission. There will be job creation in the installation and maintenance of this equipment and there will be job growth in the manufacture of the equipment. But once the new physical infrastructure has been created, less labour will be required to maintain it. Clerical workers will be displaced and women will be particularly affected. Mechanical skills will be less important than electronic and computer skills.

Retailing

Another area of service employment which has grown substantially in recent decades is that of retail distribution and trade. The number of retail outlets has decreased with the growth of supermarkets, and this has been accompanied by the trend towards self-service stores and increased economies of scale in larger stores. There has been a trend towards more part-time employment for women in retailing recently which reflects the extension of self-service forms of retailing from larger to smaller retail outlets, the continuing shift of trade toward large stores, as well as the use of computers for stocktaking, electronic cash registers, etc. Future technological developments include the laser screening of bar-coded price tags, electronic fund transfer from bank accounts and increased efficiency resulting from automated warehousing and stock control. A recent study carried out in the UK suggests that there will be a decrease in sales representatives, and increased demand for computer hardware skills; a net substantial decrease in retail staff, with an increasing emphasis on social skills. Retailing is also expected to be a low-paid occupation. The diffusion rate of all these technological changes that are expected in terms of competition, mergers, the capital costs involved, etc., is uncertain. However, the future employment prospects overall are far from encouraging.

The Leisure Industries

Another word which has become increasingly common in the debate on the future of society under microelectronics is the notion of the 'leisure society'. Here a picture is painted of increased demand in leisure-related activities such as holidays, entertainment, adult education, personal service of various kinds, recreation and craft activities. Since many of these service activities are labour – rather than capital-intensive, they have been identified as potential growth areas for employment creation.

It is difficult to arrive at any predictions about the possible job creation potential of these areas because the eventual shape of occupational structure in this part of the service sector depends so much on future government policy and the economic climate. It is misleading to portray a future where individuals are free to indulge in a wide range of leisure activities without taking into account a large number of social, political and economic constraints.

There has been very little research carried out in relation to possible job creation in the leisure area. Certainly there is likely to be an expansion of jobs in the entertainment sector as home-based entertainment becomes increasingly diversified. There is also a trend towards more self-sufficiency in repairs and maintenance of household gadgets and in consumer hardware. A continued growth in the 'do-it-yourself' industry can be expected, but the creation of jobs resulting from this is uncertain.

There is much more research relating to travel, hotels and catering generally. The changes that have taken place in the prosperous industrialized countries in the post-war years in this area are too well known to need stating in detail here, and this sector has grown rapidly in the past three decades. In hotels and catering, for example, well over a million people were employed in Europe in the early 1980s – more than 1 per cent of the total work-force. Many more people were employed in tourist-related occupations. So far little use has been made of microelectronic technology in the hotel and catering industry but there is ample scope for introducing computers for billing, stock-keeping and organizing wages and accounts in the larger hotels and restaurants. Large productivity improvements are possible by introducing self-checking-in facilities and expanding customer services by links with telecommunication information networks giving local information and providing means for booking referrals, etc. but the scope for this is limited by the need to provide a personalized service.

Clearly the future of the hotel and catering industry is so much dependent on the economic climate with the effects this will have on tourist and business travel. We can expect slow growth in employment in this sector during the 1980s, largely resulting from increased consumer expectations in the quality and range of entertainment and catering services provided.

Government Employment

Government employment deserves detailed consideration as a possible means of employment creation. This sector is seen by many optimists as the best hope for employment creation – especially in the so-called 'caring' industries, such as health and education. What are the prospects here?

This area of public employment, or the 'non-marketed' services sector, has expanded substantially in the post-war years as Keynesian demand-management economic policies have been pursued by governments in the Western industrialized nations. In 1975 19.3 per cent of total civilian employment in Britain was located in public education and in the health service compared to 12.6 per cent in France, 16.1 per cent in the Netherlands and 14.0 per cent in Italy. This increase in government employment, which has been underpinned by the growth in the welfare state, provided significant employment opportunities for clerical workers, part-time workers and women.

Recently, however, the whole philosophy of the welfare state has been questioned, and the latter half of the 1970s saw monetarist economic policies being adopted by governments throughout Western industrialized nations in response to lower rates of economic growth, higher inflation, budget deficits and the rise in oil prices in 1973-4.

It is difficult to avoid the political and economic issues surrounding the welfare state when considering the likely impact of new technology on employment prospects in this area. For example, the new microelectronic technology first started to be introduced when the rate of growth of welfare began to be checked.

Claus Offe,[12] a leading German social theorist, has outlined three possible scenarios for the future of government services. The first involved a prospect of a move towards privatization, as a result of political pressure from industrialists and bourgeois middle-class interests. His second scenario was a corporatist social democratic consensus, whereby large interest groups such as trade unions, employers' organizations, consumer groups and government would seek to establish a level of social services based on a 'social contract' compatible with the economic and political interests of these interest groups by means of prices and incomes policies, the notion of a social wage and some form of economic planning by state agencies on which the large interest groups would be represented. The third scenario (which is possibly much more problematic) involved an alliance between middle-class ideas of self-reliance and egalitarianism and traditional working-class organizations. Whichever of these scenarios prove to be correct could, of course, affect the way that new technology is implemented in the government services sector.

Work in this sector is largely characterized by record-keeping, administration and a great deal of information storage, processing, retrieval and transmission. Moreover, it is not difficult to see how much of this information traffic could be carried by telecommunications – thus leading to the possible decline in the need for local branches and agencies, at least in the long term.

In the *education* sector information technology has already been introduced at all levels. Many people may have seen very young Japanese children programming toy robots, and computers are being introduced to pupils in schools in a large number of countries. Many children already have access to a personal computer at home and computer games have caught their imagination. Some universities in the United States (many of which are in the private sector) now require prospective entrants to possess their own personal computer before they can be considered for admission.

However, whilst there are many possibilities of introducing computers into schools, particularly in the form of programmed learning for spelling, arithmetic and various forms of rote learning, the employment effects will be as much the result of demographic changes and government policies on teaching-staff levels as a consequence of the technological changes themselves. Of course, we can expect to see dramatic changes in educational technology – especially in secondary and further education. Databanks, 'electronic blackboards', video recordings of all kinds of material and language laboratories will be commonplace. There may well be a shift towards home-based university education on the British Open University model. But the educational value of all these technological innovations in education is far from clear.

As Alan Burns,[13] a British computer scientist, stated:

> Behind Marshall McLuhan's phrase The Medium is the Message, lies the belief that in any process of learning, what we actually experience is not what is being taught but how it is being done; what comes across is knowledge rather than its content . . . if this is indeed the case then the consequences of replacing a teacher with a VDU link are far from clear.

Thus the message conveyed to the student is heavily influenced by the medium used in the learning process. What is learnt from books is totally different from that learnt through lectures, films, television, etc. We are still not sure of the effect that VDU learning has on the mind of the student. Is he or she more likely to believe a VDU educational package than a human teacher? To what extent can knowledge be formally structured?

As to the long-term employment effects in the education sector, we might well see a much greater reduction in staffing when 'fifth-generation' computers become widely available. This is, of course quite separate from the inevitable employment losses in the educational administration field.

The *health* sector[14] is a good example of an area where substantial technological innovation is possible. Much publicity has already been given to computer-aided diagnosis and automated screening, and there is ample scope for ADP-based systems to be widely adopted in large medical establishments to aid record-keeping and general administration. It is already technologically possible for a system of home-based medical consultation to take place and a system of self-certification for the first few days of absence from work was recently adopted in Britain. However, the trend towards self-diagnosis in medicine will not necessarily reduce the demand for doctors' services if only because of the doctor-patient relationship – which is valued so highly by both parties. There may well be a growth in private medicine for the affluent and more emphasis might well be placed on self-help groups and preventative medicine. Coupled with the use of centralized information services and databanks, this could reduce doctors' workloads.

What then, are the overall prospects for employment trends in the health sector? As Gershuny and Miles[15] state:

> A summary of overall employment trends is thus difficult to make. It depends in part upon decisions concerning the relative weight to be given to different aspects of medical provision practice e.g. kidney machines v. geriatric services; home delivery v. hospital birth and in part upon demarcation issues among the old and the new professions in health services. Given that the whole area is subject

> to state policy, restrictions in social expenditure will remain a dominant influence on employment levels. Short of social crisis, the likely outcome is a slowdown of growth with no sharp breaks in occupation trends.

The remaining areas of government service employment involve social security, tax collection, and other miscellaneous services. We have already seen that the scope for employment displacement in administration is immense. Several EEC countries including Britain and the Netherlands are already considering a move towards separate assessment in tax collection, and such self-assessment could even be used to calculate social security claims, provided that adequate control systems could be implemented to insure against fraud.

There are many government service occupations that are unlikely to be affected much by new technology, such as refuse collection, roadsweeping, security work, parks and gardening maintenance and school caretakers. Again, employment levels associated with these occupations will be highly dependent on political decisions concerning whether or not they should be privatized, and the degree to which such services should be administered by central or local government. Unless such occupations are deliberately singled out as a means of creating employment in the form of community work, they will hardly provide many new job opportunities.

Conclusions

The prospects for continued employment growth in the existing service industries look bleak. There seems little likelihood of the service industries being able to provide new job opportunities to accommodate those people who are displaced from other sectors or who are entering the labour market for the first time. So the unemployment threat posed by information technology still looms on the horizon. What possibilities are there for jobs in the future?

Gershuny and Miles have pointed out that models of economic and technological development in society and the changing modes of service provision associated with them have to be considered in relation to the 'appropriate material infrastructure' as well as the availability of new sorts of consumer equipment:[16]

> We cannot use cars to produce transport services until we have a road network; at an early stage in the diffusion of the new 'self-service' mode of transport provision, the building of roads generated a considerable potential for the expansion of the market for motor cars. Once the social innovation has completely diffused through the society, however, the market for cars is effectively saturated, and at this point, though there may be a need for new roads to reduce congestion, the building of a new road system does not provide the same range of opportunities for expansion elsewhere in the economy.

Thus the old infrastructural provisions such as the road network, the electricity supply system, the telephone network and the broadcasting system may all require some new investment but not of a scale commensurate with a radically new form of social innovation. They see the developments in information technology as providing possibilities for a wide range of innovative modes of service provision, particularly in areas which have not been subject to social innovation in the past.

Unfortunately, information technology so far has been seen as a means of enhancing *process* innovations (that is, changes which increase the efficiency of production for existing markets) rather than as a basis for *products* with substantial new final markets.

At least one of the reasons why the new information technology is not being used to create products which are linked to a new form of social innovation is that there is as yet no appropriate infrastructure:[17]

> It may be that the telematics technologies can no more be used for the household production of information-based services without an appropriate telecommunications infrastructure than cars could produce domestic transport services without a road system. In short, economic growth depends on social innovation which in turn depends on the provision of appropriate infrastructure.

Thus one possibility of creating new job opportunities in the service sector of the economy in the future is to seek to achieve economic growth by investing in a new 'telematic' network. All this would require new patterns of legal and public control of the new network and questions of ownership, property rights, patents, etc., would have to be decided. This infrastructure would enable new types of consumer services to be provided to households, enable people to gain new skills, and provide opportunities for communal production on a local basis where the distinction between paid and unpaid employment would be increasingly blurred. There may even be a need for some form of system which rewards people not in monetary income but in the form of a trade-off between *leisure* time and *work* time. We will return to this idea in chapter 7.

We have seen that the new technology poses a very real threat of widespread unemployment. There is a possibility that the service sector may create some new jobs but it is clear that this is by no means certain because whilst substantial investment in a new telematic infrastructure is the prerequisite for economic growth it does not necessarily follow that either economic growth or net job creation will result.

We must therefore conclude this chapter on a note of uncertainty. The 'post-industrial society' or 'information economy' that is just around the corner will not take on any predetermined form. Rather, it will present us with a wide range of *alternatives* – possibly involving new ways of considering the relationship between paid employment and leisure.

5

The Trade Union Response

The history of trade unionism is strongly linked to the way technology has progressed from the Industrial Revolution onwards. The first industrial revolution, concerned with the development of steam engines, was followed by a series of secondary revolutions – those of the railway, electricity and electronics. The evolution of technological progress has shaped the labour movement, and it is in the history of the labour movement that the discontinuous nature of technological change can be seen most clearly. The very essence of trade union history is one of alternating advance and retreat; of gains secured, eroded and restored; of organizations formed, degenerating and reformed.

It is a history which can, perhaps, best be portrayed in a 'Darwinistic' pattern. Here, trade unions are seen as life-forms which continually have to adapt to threats posed by changes in the economic, political and technological environment. Each successive period of economic advance has presented contradictory faces to labour, undermining the old framework of economic security and, at the same time, creating opportunities for new and enhanced forms of labour influence. With the development of the steam engine, working people were split into two groups. While the new textile factories undermined the economic position of the weavers and stockingers, the same process enabled the millwrights to develop craft unions and pioneer local collective bargaining with their employers. Eventually, these occupations were to be absorbed into national unions following the attempts of the textile machinery manufacturers to smash the embryonic unions in a lockout over control of the machines in 1851.

In 1868, the craft unions in Britain set up the Trades Union Congress as a body to campaign for more favourable legislation to permit trade unions to carry out their activities. By the 1880s, trade unionism had developed on a national scale, and organization had spread to unskilled workers in a large number of industries. By the end of the nineteenth century, millions of people had moved to the cities as unskilled workers to man the infrastructure of transport and utilities which underpinned large-scale industrialization.

A characteristic feature of trade unionism in the twentieth century has been the growth of white-collar trade unionism, reflecting the growth of science-based industry and office employment.

At the risk of over-simplification, each phase of technological development has been accompanied by a shift in the locus of industrial bargaining power; firstly from the home to the factory tool room, then from the tool room to the mass production line, and finally to more remote control and planning administration. Throughout these phases of technological development, new occupations have come into being and either old occupations have disappeared or the bargaining power associated with them has been eroded.

As we saw in earlier chapters, trade unions have always been conscious of the impact of new technology on their jobs and skills, but their fears of widespread job displacement were never realized because new industries were created and fresh employment opportunities arose, particularly during the 1950s and 1960s. During this period, rapid technological change in Europe was managed with a certain degree of success against a background of high levels of economic growth and low unemployment. In contrast to the United States, where at that time a debate was raging over whether automation led to unemployment, the European unions generally took a positive view of technological change, provided that it was against a background of growth, and benefits could be obtained through higher wages. In many European countries, productivity agreements were signed which were specifically designed to promote technological change.

The British Trades Union Congress, following a survey of the effects of the early computers on employment levels during the 1960s,[1] felt that there was no intrinsic reason why technical change should cause large-scale unemployment. Contrasting the position in 1970 with that in 1870, the report stated:

> Any trade unionist of that time with the ability to foresee the technological changes that have taken place in the past century might well have feared that all the goods and services produced a hundred years ago could have been supplied with modern techniques, by a fraction of the 1870 labour force. In fact scientific and technological

> change has made many new products possible and our requirements have increased at least as fast as our industries' ability to meet them.

However, in the 1970s, trade union concern began to mount (particularly in the Western European countries) at some of the potentially negative effects of technological change. This was stimulated by two factors; first, the growth of unemployment and the decline in rates of economic growth, and second, the increasing concern about the effects of the new technology on the quality of working life and the potential effect of the new technology on working conditions. Significantly, this was in marked contrast to the United States and Japan, where there was little concern shown about the employment implications of the new technology. In part, this can be explained by the fact that the mid-1970s were periods of relatively rapid growth in both countries. Both Japan and the USA gained from some of the shifts in the pattern of world production and employment generation that were taking place, and against a rapid background of relatively rapid growth, the traditional areas of highly unionized labour were less seriously affected by what job losses there were.

By the end of the 1970s, as the effects of unemployment and the recession were being felt throughout the Western industrialized countries, even the labour movement in the United States became concerned about the employment effects generated by new technology.

Surrounding the renewed concern of unions in the United States about the threat posed by new technology are the rapid changes that are taking place in the location of industries in the USA; these changes are already having adverse effects on trade unionism generally. As Markley Roberts[2] points out:

> As job content changes, so does job location. The number of manufacturing jobs in the industrial North dropped between 1966 and 1977 – down 12 percent in New England, 19 percent in Middle Atlantic States and 7 percent in Great Lake States – while growing 18 percent in the Southeast, 40 percent in the Southwest, 39 percent in the Rocky Mountain States, 10 percent in the Plains States and 8 percent in the Far West. This suggests that new plants with new technology are in historically non-union or anti-union states, while some areas of traditional union strength are left with aging, obsolescent plants and equipment. Technology transfers from U.S. multinational corporations to other nations also have drastic effects on job opportunities for American workers.

In 1973 the *Washington Post* introduced computerized typesetting without union agreement. During the ensuing industrial dispute, 25 managers operated the computerized system, taking over the jobs of 125 typesetters. As the United Auto Workers described this dispute in one of their news-sheets in 1979:

> If, (at the beginning of the 1970s) someone had told workers at the Washington Post that new technology would rob them of their jobs, eliminate their skills, or destroy the power of their union, the workers, especially the highly skilled printers and pressmen, would have laughed at, or ignored, such a ridiculous idea. Yet by the end of 1973, management at the Washington Post had totally destroyed the union, taken away many benefits, and eliminated a number of jobs – all almost solely through the use of new technology. A similar story can be told of countless newspapers across the country.

This dispute was one of the first indicators of the potency of the new technology, and it was to be the forerunner of several disputes stemming from new technology in subsequent years.

Technology and the New Economic Context

Despite the fact that trade unions are able, perhaps for the first time in their history, to produce well-researched, considered policy statements to react to a major technology which is still in its embryonic stages, the ability of the trade unions to observe and anticipate future developments has to be seen against their inability to influence the direction that new technology takes.

There are obvious reasons why this should be the case: the emergence of microelectronics in a period of economic depression; the dominance of multinational companies and national governments in controlling and exploiting the technology; the fact that the technology itself has a particular form and design before it is presented to the work-force as a *fait accompli*; and the rapid pace of technological development itself. In such circumstances the criteria of profitability, efficiency, and international competitiveness which underpin technological innovation have taken on a momentum of their own. In this bleak economic and political climate the most rapid restructuring of work ever undertaken was taking place.

It is important to bear this context in mind because the trade union response to change is inevitably conditioned by the state of the labour market. During the 1970s the labour market showed a steadily rising labour surplus, and it was to be expected that the trade union response to the new technology would be subsumed into the overriding aim of trade unions to better their position in relation to both employers and government by seeking to tighten up on the labour market. In other words, by seeking to reduce the length of time worked (reducing labour supply) and/or arguing for reflation (increasing labour demand).

The Emergence of New Technology Agreements

By the beginning of the 1980s, two main strategies had evolved for responding to the implications of technological change. The American unions attempted to build outwards from the system of corporate (or company) bargaining by seeking to establish global collective bargaining with the large multinational companies such as Ford, Massey Ferguson, IBM and Philips by the setting up of international joint union councils under the auspices of the international trade centres. This strategy only achieved limited success, however, because such bodies only achieved very limited measures of recognition from the companies, and what little success they had was largely concerned with the synchronization of pay settlement dates. The Western European unions sought to persuade their national governments to reflate the European economies and to exercise state power as a means of regulating the large multinational companies, and to seek to control the direction of the new technology by a combination of data agreements and new technology agreements with companies at plant and company level.

Data agreements were pioneered originally by the Scandinavian labour movement at the beginning of the 1970s, and they were later to become the model for many of their sister European trade union movements to follow. In chapter 6 special attention is given to developments in Scandinavia, because it is there that the unions have enjoyed most success not only in negotiating data and new technology agreements but also in persuading the Scandinavian governments to promote favourable legislation to enable the unions to exercise an influence over the direction of new technology.

In Britain, the TUC began its response to the developments in microelectronics rather late. The 1978 Congress had requested the General Council

> to carry out as a high priority a comprehensive study of the employment and social consequences of advances in the new microelectronic technology and similar advances in UK technology, together with the wider ramifications of its applications by our competitors.

Congress went on to call for the General Council to prepare a draft policy statement in consultation with affiliated unions for a conference on the subject: the matter would be further taken up at the 1979 Congress. Following this, an interim report[3] was produced by the General Council as a basis for discussion at a special conference of affiliated unions on employment and technology held in May 1979, which was attended by 116 delegates from 61 unions, representing 90 per cent of all affiliated membership. Afterwards, another report, which was an amended version of the interim report, was produced for the September 1979 Congress. This second report[4] formed the basis for a lengthy composite motion moved by a number of the largest unions. The report and motion, which gained widespread support from affiliated unions, offered the most substantial and coherent statement from organized labour on new technology since the 1955 TUC debate on automation.

Many of the concerns expressed by the unions in 1955 re-surfaced, and many of the arguments expressed remained fundamentally the same. There were, however, two major differences between the two debates. The first was that the context surrounding the two debates was radically different. The 1955 debate took place when trade union members had experienced no job losses which could be traced to computerization, and the future potential job loss that could be expected was largely based on guesswork. In contrast, the second debate took place when it was becoming clear that the capacity of automation to destroy large numbers of jobs was immense. The second difference was that in the 1979 debate there was an emphasis placed on the need for government planning and intervention, and on the explicit linking of new technology with industrial democracy and planning at company and national level.

The TUC report on *Employment and Technology* stood at the peak of a serious and substantial attempt to introduce new technology and the concept of bargaining to a very large number of officials and ordinary members.[5] The report began by stating that: 'whether technology will prove to be a friend or foe will depend, not on the technology itself, but on the application and the policies adopted by governments, trade unions and employers'.[6] It then went on to argue that there was a correlation between high productivity growth and low unemployment rates, and argued for an expansion of the services sector of the economy and of R & D effort. There then followed a brief explanation of the development of microelectronics and its applications before embarking on its main task, which was to discuss the possible employment effects and to formulate a policy for the trade union movement to adhere to in order to cope with the challenge of the new technology.

Undoubtedly, the most important policy recommendation of the report was that *'new technology agreements'* (NTAs) should be negotiated wherever possible. A framework for these agreements, which was set out as a 'checklist for negotiators' formed the last chapter of the report. The 10-point checklist is set out below:

1. *Objective of 'change by agreement'*: no new technology to be introduced unilaterally; status quo provisions recommended.
2. *Challenge to union organization*: inter-union collaboration in negotiations; build-up of technical expertise by unions; technology stewards.
3. *Access to information*: all relevant information to be provided to union representatives before decisions taken; linked to regular consultations on company plans.
4. *Employment and output plans*: preferably no redundancy agreements, or improved redundancy payments if impossible; planned approach to redeployment and relocation of workers; pursue commitment to expanding output.
5. *Retraining*: provision of retraining, priority for those directly affected by new technology; principle of maintained or improved earnings during retraining.
6. *Working hours*: scope for reducing working hours and systematic overtime.
7. *Pay structures*: avoid disruption to pay structures and polarization of work-force; ensure income levels maintained and improved; move towards single status and equal conditions.
8. *Control over work*: union influence over systems design

and programming; no computer-gathered information to be used in work performance measurement.

9. *Health and safety*: stringent standards for new machinery and processes including visual display units (VDUs).

10. *Review procedure*: joint union/management study teams to monitor developments and review progress.

Given the more hostile economic and political climate at the end of the 1970s, it was not surprising that the British trade union movement should adopt a response to the new technology which could be characterized as adaptive and retrospective – very much in line with past responses to earlier technological innovation. Thus, given the *raison d'être* of British trade unions as being primarily concerned with collective bargaining both as a means of maintaining and improving the terms and conditions of employment of their members and of limiting managerial prerogative, the approach was essentially defensive, concentrating on job losses, pay and working conditions. The fundamental questions about the wider societal impact of the new technology was largely subsumed under a broad acceptance of technological inevitabilism. This response can perhaps best be summarized by the introductory paragraphs of the report:[7]

> Technological change and the microelectronic revolution are a challenge, but also an opportunity. There is the challenge that the rapid introduction of new processes and work organization will lead to the loss of many more jobs and growing social dislocation. Equally, however, there is the realization that new technologies also offer great opportunities – not just for increasing the competitiveness of British industry but for increasing the quality of working life and for providing new benefits to working people.

Significantly, the TUC failed in its attempts to secure an understanding with the employers' organization, the CBI. The decision to attempt to negotiate such an understanding was taken at the January 1980 meeting of the National Economic Development Council (NEDC). The meeting considered the TUC's *Employment and Technology* document and the CBI's *Jobs: Facing the Future*, and it was felt that there was some common ground. The TUC had a lot more to gain from such an understanding than did the CBI. The recession was biting hard, and the employers had considerably more bargaining power as unemployment continued to rise. An umbrella code of practice setting agreed standards, and guaranteeing consultation, would clearly set a valuable precedent for company and plant agreements. The TUC also hoped that after striking a concord with the CBI on new technology it could then proceed to broader areas of agreement on matters such as trade and investment, pricing and even incomes. However, the CBI, after the discussions had reached an advanced stage, decided that it was not prepared to be used as a pawn in the TUC's anti-government stance. While both sides could agree on relatively modest measures such as the need for more effective manpower planning and training, the provision of information, extra payments for extra skills and consultations on health and safety, the bulk of the CBI membership felt that a national agreement on technology would be a nuisance, and not something necessary to concede.

The TUC philosophy also encountered several criticisms from several radical commentators. For example, Mike Cooley, an AUEW - TASS past-president and the main author of the Lucas Aerospace combine committee's alternative corporate plan, argued that the TUC had implicitly assumed that technology was neutral, whereas in fact it was not.[8] This criticism was also made by Manwaring,[9] who argued that:

> The technology is not socially neutral, but embodies, and is developed within, antagonistic relations of production . . . the silicon dream is of increased leisure and higher material standards of living brought about by shop-floor bargaining over the introduction of new technology and sympathetic government action . . . . however, the critique above suggests that the dream is more likely to become a nightmare.

Whatever one's view about the neutrality or otherwise of the new technology, there were a number of contradictions in the TUC approach. First, whilst the TUC (and nearly all its affiliated unions) saw the shorter working week as a universal palliative to cope with the impact of new technology, this belief has to be reconciled with the fact that employers have maintained a hostile stance to any major reduction in the length of the working week. Moreover, it is debatable (and doubtful) whether labour markets work in such a way to ensure that a reduction in the length of the standard working week necessarily saves jobs. As we will see later in this chapter, the British trade unions have singularly failed to achieve such reductions in the new technology agreements that have been negotiated so far.

Second, the 'social democratic' or 'Keynesian' philosophy which underlaid the commitment of the TUC to relying on future economic growth as a means of facilitating the introduction of the new technology must be set against the subsequent reality of the continuing recession, the long-term decline of the British economy, the demise of Keynesian demand-management policies on the part of government, and the absence of any coherent planning mechanism to usher in the new technology with the minimum of economic and social disruption.

Finally, the organization of the British trade union movement itself, with its complex structure of overlapping and interlinking forms of organization cutting across occupations, skills and industrial boundaries does not lend itself readily to coping with microelectronic technology. The standard pattern of multi-unionism at plant level means that it is difficult to achieve a common approach to a technology which blurs occupational and skill boundaries at the work-place.

A Review of New Technology Agreements in Practice

Given the enormous effort put in to formulating a coherent policy by the TUC for British trade unionists, how do the new technology agreements that have been negotiated so far compare with the 'model' agreements suggested by the TUC? In a small survey of new technology agreements carried out by Manwaring[10] based on information reported in issues of *Industrial Relations Review and Report*, a number of agreements which had been negotiated by trade unions in 1979 and 1980 were analysed. According to Manwaring:

> The most striking feature of these agreements is that unions have not secured a reduction in working hours. In fact, none of those presented have seen hours cut. . . . Similarly, though unions have often – though not always – secured agreements that there will be no reduction in earnings and no downgrading for those offered redeployment, they have seldom secured an increase in earnings for those operating new technology. . . . In short, new technology does not guarantee increased earnings and there remains much which must be bargained for.

Whilst there was evidence in this study that the unions had been able to achieve agreements ensuring no compulsory redundancies, such guarantees merely protect the job security of the *job holder* but they do not preserve the *job position*. There was some evidence that certain unions had secured agreement on the following: that computer systems will not be used to monitor work performance; that there will be no subcontracting of work; that there will be no unilateral introduction of shift work; that work will not be deskilled; that there will be training and attendance payments for training courses; that there will be a union membership agreement (closed shop). However, these achievements won by the unions tended to be the exception rather than the rule, and as Manwaring states:

> It would, therefore, not be an over-exaggeration to conclude that unions have been largely unsuccessful in securing a share of the benefits of new technology: the rhetoric of model agreements has not, in general, been translated into negotiated concessions in clauses of actual new technology agreements.

A much more comprehensive analysis of 100 NTAs, many of which were signed in 1981, was carried out by Williams and Moseley of the Technology Policy Unit at the University of Aston.[11] The authors defined NTAs as 'formal agreements which directly and explicitly attempt to exert trade union control and influence over the process of technological change and its effect on work and conditions of employment with new technology'.

An important finding from this study was that over 90 per cent of the agreements that were signed were exclusively concerned with white-collar workers. Moreover, 75 per cent of the agreements had only one union signatory to them, despite the fact that new production and administration systems frequently embrace a wide range of occupations in the work-place. This finding is particularly significant because trade union studies which were carried out in British Leyland (Cowley) by the Trade Union Research Unit at Ruskin College, Oxford and studies produced by the 'Coventry Workshop' of GEC Telecommunication and Alfred Herbert in Coventry, highlighted the detrimental effect that new technology had on trade union organization in the work-place.[12] In particular, the researchers in the Coventry studies concluded that collective bargaining was not an effective mechanism for negotiating technical change. Problems are found in union organization, the overloaded agenda of wage negotiations, and in the lack of information disclosure and trade union training.

In addition, the Coventry trade union study showed that the differences between groups of employees (skilled and unskilled, technicians, designers, foremen, computer specialists, clerical workers, management) were considerably sharpened. All of these groups perceived the threat of new technology differently, and since each group was represented by a different union, inter-union relations were soured as a result. It does seem, therefore, that even when trade unions in Britain *do* manage to persuade management to negotiate a new technology agreement, it is difficult for them to bury their inter-union and inter-occupational differences.

One important feature of NTAs that are of considerable value to trade unions (if they are able to achieve it) is a provision which allows them to influence technological change by being involved in management's proposals from a very early stage. Trade unions have much more to gain from taking the initiative with new technology and hence shaping it from the start than in reacting at a late stage to plans presented to them which have already been formulated by management. The Williams/Moseley study examined the content of the agreements and classified them according to a range of procedures which employers undertook to follow with the unions if they wished to introduce such technological change. Overall, the author found that 'mutuality' was an exceptional achievement, although a majority of agreements stipulated that negotiations should take place before change occurred.

In terms of access to information on technological change, the Williams/Moseley study found that on a number of specific topics (type of equipment, proposed siting and timing, extension of equipment, work methods, work flow, manning, skills, training, remuneration, effects on careers, effects on job satisfaction, health hazards, and costings), over one-third of all NTAs contained no clauses at all relating to the provision of information.

One of the objectives that the TUC were seeking to achieve in the 'Check-list for Negotiators' which was published in its *Employment and Technology* policy document was to encourage unions to ensure that income levels were maintained and improved, and that increases should be shared out amongst the whole work-force to avoid a polarization between a highly skilled and highly paid elite and an unskilled, low-paid work-force. The Aston study found that only 13 of the agreements they examined explicitly linked the introduction of new technology to shorter hours and higher pay. The researchers concluded that the TUC argument that 'the case for accepting technological changes rests largely on a fair distribution of the consequent benefits' was far from being achieved. In almost all cases where pay was increased for accepting new technology, such increases were confined to new technology operators themselves rather than being distributed throughout the work-force as a whole. This may, of course, have been no more than a reflection of the drastic decrease in trade union bargaining power as a result of the recession.

Given the widespread predictions of technologically induced job loss, what did the agreements specify on questions of manning levels and grading? Only a small number of agreements included specific commitments to at least maintain the current levels of manning (4 per cent) and grading (10 per cent), and the most common approach was one of protecting existing job holders at the expense of people not employed in the establishments. Thus 36 per cent of agreements dealt with manning reductions by natural wastage, with a further 14 per cent allowing for voluntary redundancies as well. 39 per cent of agreements stipulated that an individual's wages/grading would be maintained, even though the *job might be downgraded* (that is, replacement labour would be at a lower grade); 12 per cent of the agreements even set up mechanisms which allowed *downgradings* of individuals (typically after a fixed period of time had elapsed, or following consultation on, for example, redeployment).

More ominously, the authors had noticed a distinct *tailing off* of NTAs in the 12-month period of their study and there was much evidence to suggest that most technological change was being introduced unilaterally by management without any kind of agreement at all. Even where an NTA is in operation in an establishment, if it is not adequately monitored it can be used as a facilitative device by management, thus removing the need for any renegotiation when new technological systems are introduced in the future.

Health and safety is the most constant and by far the most detailed element in the new technology agreements that have so far been signed. This may be because there is already a great deal of legislation in Britain (in particular, the Health and Safety at Work Act, 1974) which brought into existence health and safety representatives who had already built up substantial experience by the time new technology started to be introduced (particularly equipment such as VDUs). In any case, negotiations over quantifiable standards relating to the construction and installation of VDU equipment (see Box 5.1) gives much scope for precise bargaining. Certainly, when the early VDUs were first introduced they were often installed into environments which were badly lit, poorly ventilated and ergonomically deficient. Other union concerns related to complaints which were often expressed by operators that VDUs caused eye strain and fatigue. Since then there have been significant improvements in VDU design by manufacturers and the issue of VDU health hazards is no longer as important as it once was. Indeed, the Aston researchers found that most of codes and standards incorporated into VDU safety clauses in the agreements they examined had been lifted wholesale from union guidance documents!

BOX 5.1

Swedish Regulations on work with VDUs

The Swedish National Board of Occupational Safety has published Directive No. 136 on *Reading of Display Screens* which gives the following rules for operation:

1. Ambient lighting must be suitably adapted. Special importance must be attached to lighting conditions at work-places where reading of display screens occurs regularly. Generally the illuminance required is lower than in ordinary office work. In work-places where work is continuously conducted at display screens, an illumination of between 200 and 300 lux may be suitable.
Note: Lower illumination levels may be appropriate in certain working environments of a special nature (for example monitoring and traffic control).
2. When ambient lighting is subdued as per point 1, supplementary lighting must be adjustable and fitted with glare control arrangements.
3. Excessive differences of luminance in the field of vision produce what is termed contrast glare. The work-place should therefore be organized in such a way that the background of the display screen is of suitable luminance and the employee's field of vision does not include a window or any other glaring luminances. Bright reflections in the display screen are to be avoided.
4. The visual distance to the display screen and the angle of inclination of the display screen should be individually adjustable with due regard to ergonomic requirements. In the case of employees who wear spectacles, it is important that the optical correction is well adapted to the visual distance and vice versa.
Note: Ordinary spectacles for private use are often unadapted to the visual distance occurring in display screen work. Traditional bifocal lenses are unsuitable in many cases, because they often entail a strenuous work posture when used for display screen reading.

5. If an employee has a refractive error and incurs visual discomfort in connection with display screen work when using spectacles intended for normal purposes, the display screen must be moved to a position where the discomfort is eliminated. If this is not possible, the employer is to provide the employee with special spectacles which have been tested for display screen work.
6. If eye fatigue or visual discomfort tends to develop, the work must be organized in such a way that the employee can intermittently be give periods of rest or work involving more conventional visual requirements.

In contrast to the emphasis which NTAs have placed on the physical environment and equipment specifications, the subject of job design and its effects on the quality of working life has received less attention.

In sum, the new technology agreements that have been so far signed in Britain have been modest, and largely limited to the white-collar sector. There seems to have been a great deal of resistance by employer, both individually and collectively, to agreements and all the evidence points to redundancies being created in many instances where new systems are introduced. It appears that the recession, rising unemployment and a government committed to reductions in public spending and set against any kind of planning for industrial and economic expansion have all combined to severely limit the ability of trade unions to influence the direction and form of the technology by means of new technology agreements. Indeed, it may well be the case that NTAs are now less common than they once were in the immediate post-1979 period, and where they do exist some of them are seen by management as facilitative devices which give them a *carte blanche* to introduce whatever new technological systems they may wish in the future.

Manwaring may have been right when he wrote:[13]

> The silicon dream . . . of increased leisure and higher material standards of living brought about by shop-floor bargaining over the introduction of new technology and sympathetic government action . . . is more likely to become a nightmare.

Trade Union Experience in other European Countries

France

In France the trade union movement is characterized by a low level of union membership and rival political divisions. The ability of French trade unions to bargain effectively in large areas of the economy is severely limited. In the area of new technology this weakness is compounded by the exceptional importance which has been placed on information technology–*'informatique'*–by recent French governments. The new socialist government under President Mitterand, ever since it took office in mid-1981, has moved towards the implementation of a shorter working week regulated by statute. This measure was seen as a means of combating any unemployment induced either by the recession or the introduction of new technology.

In 1980, trade union confederation the CFDT (Confédération Française Democratique du Travail) published nine propositions for the introduction of new technology. These put forward guidelines for the discussion and consultation of works councils on the subject and covered: information on investments, information on effects, early consultation, pilot experiments, the use of outside experts, time for assessment, reviews of the systems in place.

The CGT–FO (Confédération Générale du Travail – Force Ouvrière) proposed the establishment of a single body where all computer technology and industrial automation questions would be reviewed regularly by the social partners. The CGT–FO also campaigned for the government to place a higher priority on employment issues in its economic strategy; in particular, the union advocated that the government should increase its aid to the regions and investment in socially useful products.

Since the beginning of 1982 the Mittérand government have been encouraging the employers to conclude so-called solidarity contracts with the labour administration. Their goal is to create new jobs for jobless school-leavers and adult unemployed persons with large wage and social security subsidies, coupled to sensible reductions in working hours and the retirement age. The first results of these measures were published in early 1983 after 5 months of practice:

* 3,661 contracts had been signed, representing about 51,000 job openings;

* early retirement by resignation was by far the most successful formula with 3,442 contracts providing about 45,000 new jobs;
* phased early retirement and a reduction of the length of the working week came last; the reluctance from the employers' side stemmed from foreseeable problems of reorganizing the production process and of rising labour costs.[14]

There are very few examples of new technology agreements in France except in printing and in the insurance sector.

West Germany

The issue of new technology has been very much at the centre of the West German industrial relations stage for a considerable period of time. As early as 1973, members of IG Metall, the engineering union, took action over the rights of their works councils to play a major part in the 'humanization of work'. Action was also taken by German printers (members of IG Druck und Papier) in the winter of 1977 – 78 against technological change in the industry. The settlement that finally emerged contained job guarantees for skilled workers for 8 years, 6 years of which would be paid at wages comparable to those enjoyed in their former grades.[15] The system of industrial relations in Germany has a dual structure. At national and regional level collective bargaining on pay, conditions etc. takes place between employers' associations and union federations. At company or plant level, certain rights of information, consultation and joint regulation are given to works councils through 'co-determination' legislation.

At national level, agreements have been concluded across a range of major sectors which refer to the introduction of new technology. In general, these have been defensive in nature, and they have sought to protect workers from the negative effects of rationalization and downgrading. However, as long ago as 1973, an agreement was concluded in the engineering industry in North Baden-Württenberg, which also sought to cover job design questions and specified a minimum period of time for job cycles for workers on semi-automated production lines.[16]

In 1979 new technology agreements (Rationalisierungs-Schutzabkommen) in the chemical, textile and leather industries were negotiated.

They provided for:

* enlargement of employment security for older workers with long seniority in cases of technical change and rationalization;
* relatively high allowances in cases of unavoidable dismissals;
* 6 months wage and salary guarantees in the case of the need for retraining;
* 9 month's income guarantee in cases of deskilling and degrading;
* help with entitlement to unemployment benefits in connection with early retirement of 59-year-old employees.

Much of West German industrial relations is handled by the works councils, the main forum for the day-to-day practice of the co-determination principle. Technological change is one of the issues on which the councils have a legal right to negotiate, and it has been estimated that more than 60 works council agreements on the introduction of new technology have been concluded. These agreements usually contain similar clauses to those negotiated in the UK, but often with a heavier emphasis on disclosure of personal records on employees. Despite the fact that works councils lack a veto in the introduction of new technology, some significant achievements have been attained. Many VDU agreements cover issues wider than ergonomic standards covering working time, skill requirements and work organization in general.

The dual structure of industrial relations in West Germany has led to problems for the trade union movement, particularly in terms of co-ordinating the sectoral negotiating approaches, with company-level consultation. Divorced from collective bargaining, works council representatives (almost 90 per cent of all works council members are unionists and practically all council chairmen and their deputies are more than rank-and-file union members) can find it difficult to use information which is disclosed to them to effectively influence the introduction of new technology. Moreover, whilst works councils have extensive rights to consultation on issues related to new technology, there is no obligation upon the employer to go further than this in taking account of their recommendations. As they lack an ultimate sanction, they do not provide a substitute for effective local trade union organization.

Italy

The development of joint regulation of the introduction of new technology in Italy has taken place primarily through existing collective bargaining machinery rather than by means of legislation or the conclusion of new agreements specifically referring to technology. There have been a few NTAs negotiated, but these have been limited to a small number of 'high-technology' companies with highly motivated and skilled labour forces.

Whilst collective bargaining has been the main instrument of joint regulation of technological change, certain trade union collective bargaining rights are themselves derived from legislation. For example, the *Statuto dei Diritti dei Lavoratori* (the Statute of Workers' Rights) of 1970 lays down in general the right of trade unions to organize, negotiate and take industrial action at company level, but it does not apply to workers in companies employing less than 16 people, nor does it apply to government employees. Article 9 of the statute, covering work-place health and safety, gives trade union representatives the right to inspection and monitoring of work-places and consequently the right to relevant information on technical changes.[17]

There is also a Health and Safety at Work Act of 1978 which sets down rights for trade union representatives in the field of health and safety at the work-place. This statute requires the employer to supply relevant information on such things as plant, equipment, machinery and materials, but it is more concerned with hardware rather than with systems design and its implications for working conditions.

The Italian trade unions have generally preferred to handle new technology through collective bargaining agreements rather than relying on the protection of legislation. For example, although the employers are prohibited by law from degrading and deskilling their employees by being allowed to transfer them to suitable jobs only whenever any rationalization occurs, in practice this law works to the detriment of trade unionists. The reason is that workers faced with the choice of a degraded job or the unemployment queue will always prefer the former.

There have been some notable successes through collective bargaining. A 1977 agreement between Olivetti and the National Federation of Metalworkers (FLM) covered issues of investment, employment and the reorganization of manual work in the light of technological change. These agreements succeeded in establishing the principle of information, consultation and negotiation on technical change. The principle was extended to most industrial sectors in 1979 so that where an employer wishes to introduce new technology, he must negotiate with the relevant union at the appropriate level.

These agreements at national level covered six industries, and they provided extended negotiating rights on technical change, new work organization and new products for the trade unions in larger companies; for example for all metal manufacturing undertakings with more than 200 employees. These agreements covered 1.5 million employees. Other provisions were:

* the labour contracts of laid-off workers were still operative;
* the Wage Compensation Fund pays 80 per cent of former earnings levels;
* workers are obliged to accept alternative work within a 50 kilometre radius;
* after more than 5 months' unemployment the dismissed workers were entitled to retraining and further education;
* the companies concerned were obliged to re-hire all workers with more than 2 years of unemployment.

Similar arrangements were made for the chemical, textile and construction industries, as well as for the banking and trading sectors.[18]

Given the deepening economic crisis since 1979, the unions have used these agreements in a defensive way to protect their members against job losses. There have, however, been some examples of more positive attempts at trade union involvement in systems design. The experience at Olivetti was particularly interesting in that the reforms in the system of work organization during the 1970s were a result both of union pressure to decrease the monotony of certain jobs and the desire on the part of management to achieve a more flexible system of production in the light of the transfer to electronics-based products from mechanical products. The result was that integrated assembly units were introduced to replace line assembly, with the individual unit being responsible for a specified product or unit assembly. This led to an upgrading of skills and improvements in job satisfaction. This system of work organization was greatly facilitated by the modular structure of the production process of electronics-based products.

In early 1981 a new agreement was also signed with Alfa Romeo which resulted in an important step forward in replacing traditional assembly line working. The new working methods were also designed to fully exploit technological potential in the company following the agreement finalized with Nissan.

As in most other European countries, the newspaper and publishing industry was heavily involved in negotiations on the introduction of new technology. In contrast to some other countries, the Italian printing unions succeeded in obtaining a pledge from the employers that current employment levels would be maintained, and that the skills of the printers would be 'upgraded' by means of training courses in new technology. One agreement was particularly noteworthy. The agreement with the Rizzoli-Corriere della Sera group, Italy's largest newspaper and magazine group, achieved temporary suspension of the introduction of new electronic composition systems operated directly by the journalists, plus the conversion of Linotype operators into operators of the new electronic typesetting equipment. An agreement due to come into force in 1983 provided for direct input by journalists, but it stated that surplus printers would be protected by reducing overtime, increasing rest days and as a last resort, early retirement.

New Technology and Trade Unions in the United States

In the USA the trade union movement, unlike its Western European counterparts, has not developed a central trade union strategy to cope with the challenge of technological change. Much of what has been achieved has been done by individual unions as part of their normal company bargaining. Collective agreements in the United States are the most lengthy, detailed and complex in the world, and the well-developed system of grievance arbitration has generally served the needs of the unions well.

Some progress has been achieved by collective bargaining at company level. For example a 1981 Bureau of Labor Statistics study, updating a similar 1966-7 report, presents examples of labour contract language and identifies contracts which contain language on plant movement, inter-plant transfers, and relocation allowances, many of which relate to the effects of technological change.[19] Agreements limiting plant movement rose from 22 per cent in the 1966-7 survey to 36 per cent in the 1980-1 survey of some 1,600 contracts, while worker coverage rose from 38 per cent to 49 per cent. Interplant transfer provisions increased from 32 per cent to 35 per cent and worker coverage went from 47 per cent to 49 per cent. Agreements dealing with relocation allowances increased from 34 per cent to 41 per cent while worker coverage went up from 60 per cent to 65 per cent.

On the issues of major technological change, work transfer and plant closure, some contracts have a variety of provisions. For example, the UAW–General Motors contract provides for advance notice to the union in cases of technology-related permanent lay-offs, a special union–company committee to deal with technology lay-offs, and company transfer of work. In addition, this particular contract provides for retraining, preferential re-hiring based on the seniority principle and generous severance pay.

Despite the progress that has been made by US unions through company bargaining, unlike some of their Western European counterparts, they have had little success in persuading Federal or State governments to promote favourable legislation which might enable them to exercise more control over the direction and application of new technology.

Conclusions

It is clear that new technology will provide one of the biggest challenges to trade unions throughout the 1980s and beyond. It has been introduced at a time when union membership is falling rapidly as unemployment throughout the Western economies climbs steadily upwards. It is no longer possible to rely on high rates of economic growth in the economy as a means of providing alternative job opportunities in an expanding service sector. Information technology increasingly cuts across occupational boundaries and fundamentally alters skill patterns within the labour force. In such circumstances, the very heart of trade union organization at enterprise level is threatened.

So far, the early experience of workers in Britain and throughout Western Europe, confronted with rapid technological change, has been uneven. Some workers have only experienced a small amount of technological change and change has been absorbed in working practices with little dislocation. The heartlands of manufacturing industry have been most affected, with thousands of jobs disappearing in the latter part of the 1970s and early 1980s either due to radical shifts in markets or from rationalization brought about by a much more hostile competitive environment, in which technological change has featured prominently. Whilst the 'quality of working life' movement of the early 1970s offered the prospect of a move away from 'Taylorized' production lines, we now see fresh dangers that employers will seize upon their renewed bargaining power to introduce greater forms of degradation in skills and working life generally, not only on the factory floor but also in the office.

Information technology is still in its embryonic stages, but it does seem that it emphasizes the need for new approaches from the trade unions if they are to survive as a viable force. With traditional technology, the issues for negotiation focused upon the physical equipment, machinery and materials involved in the production process; in order to deal with such technology the unions were familiar with a range of standards, codes and regulations which could be used whether they were to be found in collective agreements or in legislation. With information technology, the effects upon working conditions, skills and work organization depend much more on systems design rather than on the physical hardware *per se*, and trade unions may well find that procedures and bargaining strategies which were adequate in the past are no longer as effective as they once were.

Whether the trade unions rely on the negotiation of new technology agreements or on the establishment of legislative standards, these can only be seen as the first steps in the process of seeking to regulate the new technology at company and work-place level. The crucial question is to what extent such 'codes of best practice' can be implemented in practice. Scandinavian research has suggested that the effectiveness of trade unions in exerting more influence over the direction of the new technology and the system of work organization that results from its introduction depends on a number of factors: their power base; their awareness of their potential for influencing changes; the development of work-group knowledge of the effects of the technology; the access to outside expertise; and the commitment of the trade union organization to pursue its objectives over a sustained period of time. As we will see in the next chapter, much union activity in the Scandinavian countries has been concerned with the representation of the trade union in company project groups, the creation of support groups within the union with the necessary educational back-up and the establishment of a procedure for transferring problems to a higher level.

6

Alternative Approaches to New Technology: Scandinavia

In chapter 5 we noted that the new technology agreements were originally pioneered by the Scandinavian trade unions in the early 1970s. Developments in the Scandinavian countries warrant special treatment because these countries have the most extensive rights for trade unions to negotiate on technological change in the whole of Western Europe. In part this has been achieved through the conclusion of collective agreements and in part through legislation. In addition to this, the unions in the Scandinavian countries have devoted a major effort to setting up educational support programmes for their membership at local level to enable them to make best use of these rights and to build upon these rights in the organizations where they work. As in many other European countries, the rights to negotiate have been backed up by legal regulations of standards of health and safety on the introduction and use of new technology.

It is significant that the social and employment effects of new technology have received most attention in those European countries where a continuing social 'consensus' has existed between the labour movement and the government. This is certainly true in the cases of Denmark, Norway and Sweden and, to a much smaller extent, West Germany. In each case a Social Democratic-led government has been in office for much of the past 20 years or so.[1] This has meant that although the industrial relations system in each of these countries has been based on centralized bargaining with legal restrictions on the right to strike which discouraged shop-floor bargaining, the third party to the negotiating process was a sympathetic Labour government. Not only did this mean that a continuing social consensus could be negotiated with the government by the labour movement (usually as a *quid pro quo* in exchange for co-operation in the field of incomes policy), it also meant that the labour movement could rely on favourable legislation to open up areas of management decision-making to the trade unions as well as the possibility of state support for trade union research into the effects of new technology on the quality of working life. Moreover, the Scandinavian experience on new technology can be seen as a product of a general concern on the part of the Scandinavian trade unions to 'democratize working life'. As such, many of the developments in those countries can be traced back to research which took place in Norway in the late 1960s which was commissioned by the Norwegian Iron and Metal Workers' Union, known as the 'Iron and Metal' project. This research was to lead to the negotiation of the first 'Data Agreements' – dealing with the regulation of computerized systems in the work-place – not only in Norway but also in the rest of the Scandinavian countries.

Norwegian and Swedish Data Agreements

The first Data Agreement between the Norwegian Confederation of Trade Unions (LO) (Landsorganisajonen i Norge) and the Norwegian Employers' Confederation (NAF) was signed in April 1975, and by 1982 most of the Norwegian working population was covered by these kinds of agreements.[2] This agreement was renewed, with modifications, in 1978. The agreement established the principle that 'in the design, introduction and use of computer systems, the social effects should be considered as equally important as the technical and economic effects'.[3]

The agreement lays down procedures to be followed in the introduction of computerized systems, primarily as far as the local union is concerned. It provides for the provision of information to the trade unions by management concerning proposed changes as early as possible, before decisions are made. In addition, it also requires both management and unions to provide information and involve the work-force to the greatest extent possible in the planning stage. The work-force is given the right to participate in issues stemming from the technology and to be fully trained in computer-related skills. An interesting provision of the agreement is that unions are given the right to elect a shop steward to specialize in computer-related matters, with the additional right for the 'data steward', as he/she is called, to receive full training. Another provision relates to the rules pertaining to the use, collection, and protection of computerized data held on individuals within the company. The guidelines for data collection and use should be agreed if possible at a local level, although local parties have recourse to national level if an agreement is not forthcoming.

When the agreement was renewed in 1981 for a third period, certain changes were made. The terms of reference were broadened to cover the introduction of new technology in general. There was also a change in the terminology to emphasise negotiation at local level as a means of conflict resolution.[4] The unions were also given the right to use external experts paid for by the company at local level to advise them.

This agreement has so far led to well over 700 more detailed local agreements (out of a total work-force of around 1.9 million), as well as a number of central agreements in the public sector, banking and insurance (all sectors not covered by the NAF). A particularly interesting example is an agreement with Televerket – the state-owned telecommunications corporation – in Bergen. Following health and safety problems with VDUs, a limit of 2 hours work per day with VDUs was set. In the banking sector, a section on computer-based systems was added to the co-operation agreement between the bank employers' federation and the Norwegian Bank Employees' Association. This gives more detailed specifications on the employer's obligation to provide information; it also provided for the setting up of consultation procedures and the right of representatives to participate in the planning process before new technology systems were introduced. At this stage, the management must provide a written report outlining the consequences of the proposed new technology:

a The organization of the bank:
functions that will be eliminated;
new functions;
other organizational changes;

b Employees of the bank who will be directly affected:
jobs that will be eliminated;
jobs that will be widened in scope;
changes in authority;
working out the scope of the jobs in detail;
training;
jobs that are abolished/transfers to other jobs, and opportunities for training in this connection, where special consideration is given to older employees;
the economy of the bank;
time-table showing when the existing system is to be replaced by the new one, and related training and information.

Following the implementation of a computerized system a procedure has also been set up to monitor the implementation of the agreement and any unplanned effects stemming from the technology.

During the years of operation of such new technology agreements in Norway, the trade unions have realized the importance of ensuring a close relationship between the elected representatives and the membership. At the heart of this relationship has been the development of a full education programme for the mass membership with regard to the technology. This can be traced back to the co-operation in the early 1970s between the Norwegian Computing Centre and the Iron and Metalworkers' Union in developing educational courses for union members which resulted in the publication of two textbooks for trade unionists on data-processing, planning and control. This co-operation between academics and unions was subsequently repeated in other sectors. One of the most important conclusions of these experiences was the need to encourage trade unionists at local level to fully analyse their own objectives with regard to change stemming from new technology.

Whilst these agreements are by far the most progressive that have so far been negotiated in the area of new technology, it is important to point out that they are not without inherent difficulties. As noted in earlier chapters, the new technology brings with it a much broader range of options in respect of work organization implications. Moreover, the characteristics of the industrial relations institutions (within which attempts to to improve the quality of working life are made) evolved in a period of high stability, resulting in a fairly rigid set of procedures and standards. In addition, the pervasiveness of the new technology is such that there is a real need to go beyond procedural requirements combined with rather vague goals and aspirations in order for such agreements to be really effective. The trade unions in Norway and Sweden have recognized this, and recently a new type of agreement on *'work-place and enterprise development'* has begun to emerge.

The main purpose of these agreements is to seek to influence what might be called the 'infrastructure' which determines the way new technology is *used* in working life, and, through this, to seek to exert influence on the research and design aspects of the new technology. The new agreements – which date from 1982 – are lengthy and far from precise.[5]

One such agreement between the SAF (Svenska Arbetsgivare-foreningen), the LO (Landsorganisationen i Sverige) and the PTK (Privattjanstemannakartellen) is called the agreement on 'efficiency and participation'. It contains several broad clauses, for example the statement that 'developing and improving the efficiency of the firm, together with safeguarding the employment, are matters of common interest to the company and its employees'. In relation to work organization, the agreement speaks of the concept of *continuous development* and underlines the need to make better use of the expertise, skills and experience of the employees, while ensuring that jobs, management practices and control systems are designed to stimulate and involve the employees concerned. There is a clause which requires both sides to seek means of achieving freedom and discretion in the work process. Article 4 deals with technology, and states that

> in the event of technological change, a sound job content shall be the goal together with opportunities for the employees to increase their skills and accept responsibility for their work. The knowledge of the employees should be stimulated together with their ability to co-operate with and have contacts with their colleagues.

The Swedish agreement says little about how such changes are to be brought about, and much was left to the negotiators at local level.

The Norwegian agreement on work-place and enterprise development has the following opening paragraph:

> In accordance with the goals expressed in Section 9 of the Basic Agreement on co-operation the organizations [LO and the NAF] commit themselves to joint efforts in support of developmental work at the individual undertaking. The efforts are based on the assumption that the developmental tasks will vary from one undertaking to another. The efforts must therefore concentrate on helping the undertaking to analyse its own situation and find practical working methods.

The agreement then goes on to mention several means whereby such development can best be furthered, including joint planning conferences, research grant support and the establishment of a tripartite Board to monitor the implementation of the agreement, consisting of two representatives from the NAF, two from the LO and two from the Work Research Institutes.

The main difference in philosophy between the Norwegian and Swedish agreements is that in the Norwegian case there is a much greater reliance on management/union co-operation and using tripartite institutions to oversee the implementation of the agreement than is the case in the Swedish agreement. This results from differences in experience in the two countries in the evolution of the Work Organization and Industrial Democracy programmes. Whereas Norway has experienced continuing joint management/union collaboration with a high premium being placed on the role of research to assist work and industrial democracy, the Swedish unions have preferred to make use of legislation and have been suspicious of joint management/union collaboration. This can be traced to an increasing polarization in the respective attitudes of both sides of Swedish industry throughout the 1970s[6] with unions seeking more and more legislation and employers competing with the unions by promoting alternative ideas about work organization in their so-called 'new factories' programme, based on job enrichment and advanced ideas in work design stemming from the well-known Volvo Kalmar programme in the early 1970s.[7]

Although there has been a divergence of philosophy between the Norwegian and Swedish approach to such work-place agreements on new technology, there are a number of common strands. For instance, both approaches seek to build on the *expertise and knowledge* of the workers themselves as a means of bringing some influence to bear on the way that new computerized systems are implemented. The aim here is not necessarily to try to match the resources of employers by assembling an array of 'counter-experts', but to identify and utilize workers' expertise and knowledge so that they are able to participate in technological change to their advantage. This involves the setting up of mechanisms to continually monitor new developments and to identify areas of good and bad practice; initiating and running a number of pilot projects on the democratization of working life; and establishing funding bodies to support and co-ordinate research on the social and employment implications of new technology. In the case of Norway, the development agreement between the NAF and the LO provided for 5 million Norwegian crowns (about US $700,000) per year to be allocated to research and educational programmes on work organization and industrial democracy aspects of new technology. This fund is jointly administered by the employers and trade unions. Sweden has a much larger strategic resource in the Work Environment Fund (Arbetskyddsfonden). This fund is largely financed by a levy of employers and distributes about 500 million Swedish crowns (about US$ 70 million) per year. The funding bodies in both countries, apart from promoting conventional training and education programmes in new technology and its implications for working life, are also concerned with setting up development projects to gain new insights – particularly in relation to the structure and design of technological systems so that they allow for more worker democracy and participation as well as an increase in job satisfaction and an enhancement of skills.

Of course, these agreements are not without their limitations, and the Norwegian and Swedish trade unions would be the first to acknowledge that they are only a first step towards coping with radical technological developments that are still to come. What conclusions can be drawn from experience in Scandinavia in so far as trade unions in other Western European countries are concerned?

Perhaps the most obvious point is that trade union participation in technological change is a slow and difficult process to undertake, and it has to be seen as part of a *continuing process*. It implies major policy shifts within the union hierarchy towards more involvement by the trade union membership on the shop floor and less centralized control by the union leadership. It is not enough simply to rely on hiring the services of outside consultants for advice. Rather, the Scandinavian approach seeks to build upon a broad understanding and high level of commitment on the part of the rank-and-file themselves, but co-ordinated with central support from the official union hierarchy.

Second, as mentioned earlier, trade union influence over technological change is dependent on the unions' bargaining power. This in turn is heavily reliant on having a government in office that is sympathetic towards trade union objectives. It is also partly a function of the solidarity existing between union members which can be used for collective sanctions. The pervasiveness of new technology means that trade unions have to extend their influence into areas which are outside the traditional collective bargaining agenda, and this requires an enormous educational commitment.

There are two barriers which have to be surmounted in this process. On the one hand there is the barrier presented by the attitudes of many trade unionists that automation is inevitable and that it is futile to try to 'stand in the way of technological progress'. On the other hand there is the problem of a series of isolated technical changes with limited consequences which over a long period add up to an unforeseen development of serious proportions. While externally produced information and training could, to a limited extent, overcome these difficulties, it became apparent that it was necessary for trade union consciousness to be developed out of practice by tailoring their own objectives to real experience in the work-place, rather than relying on external guidelines.

Nygaard,[8] in his review of trade union experience with the 'Iron and Metal' project in Norway, emphasizes the dangers for trade unionists in starting their education in the computer area 'by acquiring the current knowledge and understanding of data-processing of the systems workers or the managers'. If they do this, according to Sandberg,

> they will be trapped. Instead they must start by building up their own basic understanding in terms related to their job situation and the trade unions' picture of the world. Later on in the process existing knowledge may be integrated into the framework they initially establish.

Another Scandinavian author[9] found that three types of *information* were needed for effective union intervention in developments in new technology. In the first case there was information relating to the proposed system, which data agreements specified should be in non-specialist language, and should include details of the social consequences of automation. This information had to be provided early enough to enable plans to be influenced. In practice, conflicts developed over the definition of relevant information, the completeness of the material on the social consequences, and the meaning of 'understandable form'. More particularly, the information was often given to the union too late for it to influence events owing to bad management and project organization. Documentation might only be produced after a lengthy informal management process in which key decisions had already been made. A second type of information related to the collective knowledge and experience of the local union concerning the impact of new technology on working conditions and negotiating strategy. Here it was found from the Norwegian experience that the knowledge needed to be made explicit in order for it to be usefully distributed systematically through publications, seminars, etc. A third type of information related to the difficulties of finding competent outside experts who were able to adopt a trade union perspective.

Norwegian Legislation on New Technology: The 1977 Work Environment Act

Data agreements in both Norway and Sweden have been supplemented by extensive legislation which has been introduced by sympathetic Social Democratic governments, particularly so in Sweden. In 1977 the Norwegian government passed the Work Environment Act. This Act requires firms with more than five employees to elect safety delegates; firms with 50 employees or more are obliged to form *work environment committees*. The

Act gives information and consultation rights to these work environment committees on the introduction of new technology. The Act specifically covers system design and states: 'The employees and their elected union representatives will be kept informed about the systems employed for planning and executing work, and about planned changes to such systems' (section 12:83). The Act also empowers an independent government agency – the Labour Inspectorate – to issue legally binding regulations to ensure. 'the working environment is fully satisfactory as regards the safety, health and welfare of employees', and that 'technology and the work organization is designed so that employees are not exposed to undesirable physical and mental strain'.

The Inspectorate issued a set of regulations concerning the introduction and use of visual display units (VDUs), although their implementation was held up by the change of Norwegian government in the autumn of 1981. These cover information disclosure, training of VDU workers, work monitoring, ergonomic equipment standards, work pauses and sight tests. It is perhaps significant that the regulations deal more with systems design and hardware specifications than is the case in other countries where standards have been set. For example, the training requirements must have a certain content including the significance of the VDU tasks to the company as a whole; the technical make-up of the equipment; the means of making maximum use of the equipment, alternative work routines in the event of technical breakdowns; and the need to take adequate breaks and vary work. The regulations also specify that if work with a VDU is routine, operators should spend a maximum of 50 per cent of their working time at the terminal.[10]

New Technolgy in Denmark

Denmark is another Scandinavian country where developments in new technology have been very much in line with a distinctly 'Scandinavian' philosophy. Following pressure on employers by individual unions over recent years a national framework agreement on the introduction of new technology was concluded in 1981. The signatories were the Danish LO (Landsorganisationen i Danmark) and the Danish Employers' Federation DA (Dansk Arbejdsgiverforening) and the agreement, which took as a model the Norwegian agreement, covers a major part of the private sector.

The agreement supplements the 1970 DA-LO 'co-operation'

agreement on consultation on company policy. It provides general guidelines, which can be supplemented by negotiations at company level. Changes in production techniques are to be discussed at *'co-operation committees'* which already exist in most companies with more than 50 employees but, where appropriate, separate technology sub-committees should be established. Management must inform the committee before change takes place and give information which is understandable so as to enable the committee to discuss the technical, financial, social, personnel and work environment effects of the changes. The philosophy of co-operation committees is rooted in the notion that both parties 'should strive for agreement' with a commitment on both sides to ensure that the principles agreed on both sides are, in fact, applied. The 1981 New Technology supplement to the Co-operation Agreement, provides for either party to call in a special 'expert from the undertaking', and if the parties agree, to call on other 'experts'.[11] Disputes arising from the agreement are referred to the co-operation committee within the undertaking concerned, and if the matter cannot be resolved it is channelled through the central organizations to a central Co-operation Board. Arbitration at national level may result in fines for breaches of the agreement. One trade union criticism of the agreement has been that one side or the other can block the calling in of outside experts. This has limited the possibility of the Danish unions using wage-earner consultants.

There is thus a distinct difference between the Swedish and Danish approach to the problems generated by new technology. The Swedish approach emphasizes change by negotiation, in contrast to new technology issues in Denmark being subsumed under the auspices of co-operation committees. Like Norway and Sweden, the Danish unions have developed educational and training policies in support of their collective bargaining strategy. One such project is the 'DUE' project which has been established by the research council of the Danish trade unions. The project is action-orientated and involves university researchers working alongside local trade unionists in the work-place.

In addition, the Danish Working Environment Act contains a provision which entitles a safety representative to participate in the planning stages of proposed technological changes, providing that such technological changes affect the working environment.

Legislation and New Technology in Sweden

Sweden has a very centralized collective bargaining system which is based on a series of laws which provide a framework for national agreements to be concluded between central employers' and trade union organizations. In the private industrial sector, central agreements covering 1.3 million workers are concluded between the employers' organization, the SAF, the blue-collar trade union centre, the LO, and the white-collar trade union confederation covering workers in the private sector, PTK. Central agreements are also concluded for local government and health workers (0.8 million), central government (0.5 million), nationalized industries (0.2 million), co-operatives, banking and insurance. These central agreements lay down principles which are then supplemented by local agreements. Recourse to national level bargaining is normally provided for in the case of a failure to reach agreement at local level.

This centralized industrial relations system, the product of the 1938 Saltsjobäden Basic Agreement, has gained an international reputation for providing Sweden with a long period of industrial peace and stability. The consequent institutionalization of industrial conflict, backed up by Labour Courts, Labour-Management Co-operation Councils, favourable experience in collective bargaining, together with the high growth rates from the late 1930s to the early 1970s, all combined to produce the 'Swedish Model' of a stable, peaceful industrial relations system.

However, the beginning of the 1970s saw significant changes in the Swedish industrial relations climate, with increasing industrial unrest. Part of this industrial unrest was linked to issues of industrial democracy, work reform and technological change (especially increasing computerization), and this linkage was manifestly political. The trade unions expected that these issues could be partly resolved by negotiations but these would have to be preceded by substantial changes in labour law. The demands from the LO and TCO were taken up by the government, and the years 1973 to 1977 were marked by the introduction of a number of statutes designed to cover such areas as security of employment, status of shop stewards, the right to board representation, the work environment and industrial democracy.[12]

As far as new technology is concerned, the relevant legislation is largely located in the 1977 Co-determination Act and the subsequent central agreements, and the Work Environment Act of 1978 and associated agreements. However, it is important to emphasize that such legislation and agreements in Sweden have not focused specifically upon new technology as an issue, but rather have treated it alongside other aspects of management policy. The printing sector is the only sector where a specific new technology agreement has been concluded at central level. This agreement covering the newspaper industry was signed in May 1980 and covers the period up to April 1986.[13] The agreement guarantees that the introduction of new technology will not lead to dismissals. Journalists, printing workers and clerical employees would be retrained during normal working hours instead. A special education and training fund would be built up by the employers and retraining plans were to be worked out by a joint council for the industry. Under the agreement the three categories of staff would continue to carry out their jobs as before, but it would be possible for jobs to be transferred from one category of staff, and union, to another after local negotiations. Local agreements had to be approved by the national unions.

The 1977 Act of Co-determination at Work (MBL) is the most comprehensive legislation of all the Scandinavian countries seeking to extend the scope of collective bargaining to general questions of management policy. The Act covers all major questions of management policy including organizational and technological change, and requires the employer to initiate discussion and negotiate with the union *before* final decisions are taken and any changes are introduced. The employers, however, can carry out preliminary investigations without the need for consultation. This has caused certain problems for the unions who have pressed for negotiations to take place between the employer and the local union as soon as a plan for a new system is drawn up.

Another provision of the 1977 Act is that the employer is obliged to provide *information* to the trade union – on their own initiative – concerning production plans, the financial position of the firm, as well as its manpower policy. In order to obtain the necessary information about these matters the trade union has the right to examine the accounts of the company and has access to other documents concerning the operations of the company. The information must be readily available and in a form that is easily understandable.

The subsequent negotiations to translate the provisions of the 1977 Co-determination Act took 5 years for the 1.3 million private industrial sector workers, and culminated in the central agreement on *Efficiency and*

Participation between the SAF, LO and PTK in April 1982 (see earlier section on Norwegian and Swedish Data Agreements).[14]

This private sector agreement covers questions of work organization, technological developments and the financial position of the company. It provides each worker (not just union representatives) with 5 hours' paid time off per year for union activities, it allows research on working life to take place within the factories, and provides for the unions to hire wage-earner consultants *paid for by the companies concerned*. The agreement also lays down general principles of job content, education and training and information provision. While this legislation seems particularly far-reaching by European standards, it has to be seen within the context of the favourable attitude on the part of Swedish employers towards trade unionism and the value of collective bargaining. As one American commentator remarked:[15]

> The Swedish approach is determined by more than just the legal framework. Of equal importance is the widespread attitude among Swedish employers and unions alike that routine layoffs are not acceptable as a normal part of business planning. . . . There is widespread and continuing recognition throughout Swedish society of the value of labour unions in representing the interests of employees.

A key factor in the ability of the Swedish trade union movement to influence the direction of change brought about by new technology has been its ability to provide *effective servicing* at local level, in particular by the provision for the use of 'wage-earner' consultants in some of the co-determination agreements. Projects have been developed by the unions which have had the benefit of advice from systems consultants, paid for by the employers. The unions have also realized the importance of having access to research and development independently of the employers, and this has led them to extend their influence over the allocation of the national Applied Research and Development budget (particularly by making use of the Swedish Work Environment Fund, *Arbetskyddsfonden*, and under the auspices of the Swedish Centre for Working Life, *Arbetslivscentrum*).

The second area of legislation which has regulated the introduction of new technology in Sweden has been that on the working environment. A work environment joint agreement of 1976 between SAF, the LO and PTK, covering the private sector, gave local working environment committees, which had a majority of employee representatives, the responsibility for control over medical officers and safety engineers. The Working Environment Act of 1978 developed the powers of local committees and safety representatives. It extended their power to suspend hazardous work situations but also gave them the right to be consulted and an obligation to issue an opinion on any changes which can affect the work environment at the planning stage. The legislation also empowered the National Board of Occupational Safety and Health to issue regulations setting down minimum standards – for example the 1981 Regulations on work with VDUs (see Box 5.1 in previous chapter).

The Promotion of Information Technology in Sweden

Sweden has embarked on the most comprehensive policy discussions in the field of information technology, and is not only one of the leading countries in seeking to take account of the social and employment impact of information technology, but is also one of the most computerized societies in the world. An innumerable number of government and ministerial reports have been produced concerned with the development of information technology and its social consequences.

Among Swedish products, terminals occupy a prominent role – both general-purpose video display terminals and terminals designed for specific purposes such as banking. Successful Swedish applications include process control systems, industrial robots, air traffic control systems, and office terminals.

Following a series of mergers in the Swedish computer industry in the early 1970s, most of the applications of information technology have so far focused on the engineering and process industries, which have traditionally played a dominant role in the Swedish economy. The Swedish engineering industry employs nearly half the country's industrial workforce, and is knowledge-based as well as being labour-intensive. Large-scale research and development programmes are being carried out in a number of engineering fields, and computer technology is being used to create new products (such as the new generation of automatic telephone exchanges made by L. M. Ericsson). A large diffusion of advanced production technology in Swedish manufacturing industry has made Sweden the leader in the density of NC and CNC machines as well as industrial robots.[16] A recent Swedish government report envisages that CAD/CAM systems will be widespread in Sweden by the mid-1980s.

In the public sector the use of information technology has been widely applied for communications, taxation, police, social welfare purposes as well as in the civil service itself. Public concern about personal data held on individuals led to the enactment of the world's first Data Act in 1973, with the creation of a separate Data Inspection Board for monitoring and administering ADP registries and nearly all kinds of personal information in whatever form.

Future development of information technology within the civil service is expected to involve a process of decentralization of the social insurance system, and the increased computerization of office routines.

Developments Towards a Swedish Model of an Information Society

The promotion of research and development in Sweden is largely located in the role of the STU (Styrelsen för Teknisk Utveckling), a body under the auspices of the Ministry of Industry. Apart from its role of encouraging the development of products and systems and raising the level of scientific research throughout Swedish industry, the STU has identified information technology as a strategic area requiring the highest priority in resources and development. Following a 1981 report, the STU initiated a programme for 'information processing', which involved the collaboration of 400 of the largest Swedish firms and a range of institutions and research groups.

The programme involved initiatives in nearly every aspect of information processing which were allocated to a number of research groups throughout Sweden. The work is followed up, supervised and controlled by a special STU committee, with yearly visits to evaluate their scientific advances. It is intended that Sweden should have a number of scientific centres of excellence in information technology which can be used to support its long-term economic strategy.

Apart from the technical and scientific initiatives in the information technology field, there has been a serious attempt to ensure that such developments are carried out to meet the needs of users and society at large. A Data Policy Commission (Datadelegationen) was set up for this purpose by the Swedish government in March 1980 with representatives from Parliament, employers, trade unions, county councils and local authorities. The Data Policy Commission was assigned three objectives:

* to follow up developments in information technology, to stimulate competence in this field, and to propose schemes in order to guarantee the positive future use of information technology under democratic guidance and control;
* to prepare directives for the use of information technology in society;
* to support the co-operation between various government committees working in the information technology field and to co-ordinate their recommendations into an 'aggregated data policy'.

A fundamental tenet of this aggregated data policy is that Swedish information technology should be (subordinated) to major societal objectives such as

> democracy, economic growth, full employment, economic, social and cultural equality, suppression of disparities in regional development, satisfying conditions of work, co-determination in working life and equality between men and women.[17]

In March 1982 the Data Policy Commission produced a set of propositions for a 'co-ordinated data policy' which was to be further developed up until 1985. These propositions were drawn up with the purpose of creating a 'distinct *Swedish model for the development and use of information technology in society*', involving four fundamental principles: a total view is taken of all data questions; numerous individuals, institutions, and groups are given the opportunity of influencing the development and use of information technology; current problems are solved by co-operation between the interested parties; and a comprehensive educational programme on information technology and its uses should be initiated at all levels.[18]

The Work of the Swedish Centre of Working Life

As already mentioned, during the latter half of the 1970s a more critical research approach developed in Sweden in relation to the social and organizational effects of computerization. One of the central tenets of this approach was a response to the growing criticisms from trade unions on the consequences of data processing technology on the content and control of work, coupled with demands that new forms of working with computer systems should be democratized.

The Swedish Centre of Working Life (Arbetslivscentrum – ALC) was set up in 1977 as an independent body from government with a governing council of representatives from government, trade unions, and employers. It was left to the employers and trade unions themselves to decide the direction of the ALC's work within parameters which had been laid down by the Swedish Parliament.

The labour legislation reforms of the 1970s are the reference point for ALC's work. This centres on a view of working life in which the development of the employees' co-determination and participation in the decision-making process at the work-place is a way of bringing together economic and social progress. Efficiency measured in financial terms is therefore not the primary target but one that must be weighed against other legitimate interests under different conditions. The framework for this work is mainly laid down by the government in economic policy and labour legislation.

It is part of the research philosophy of the ALC that, in order to have any substantial impact on technological development, trade unions may have to propose modifications of existing technology or perhaps even develop new workable alternatives. This has to be carried out by trade union co-operation with engineers, computer specialists, linguists and social scientists. Essentially, the work is carried out by using the knowledge and experience of local trade unionists, working in collaboration with the researchers themselves.

Some Swedish Research Projects

Arbetslivscentrum has been involved in a number of research projects on working life ever since the late 1970s. One of the most well-known of these projects was the DEMOS project (Democratic Planning and Control in Working Life). The aim of the DEMOS project was to build up knowledge about the ability of trade unions to influence the planning and use of new technology within particular companies. The research method was based on the principle that trade unions, through the use of the investigative groups, should relate the work experiences and aspirations from their own members in particular enterprises in order to formulate an action programme. Four firms were included in the project and the Swedish LO and TCO, in a subsequent evaluation, cited the DEMOS project as one of the most relevant and advantageous projects for the unions.

Another example of such work is the UTOPIA project. this work involved computer specialists and social scientists working alongside each other in an attempt to develop an alternative text and picture processing system for VDU graphics. Using the existing hardware, they are developing new software that takes into account both the quality of work and the quality of the product, which is flexible enough so that it can be adapted to particular work-places. In addition, the UTOPIA project will develop training alternatives for the new system based on trade union criteria. The training is planned to take place parallel with the alternative technical systems, and will include the development and maintenance of the technology and the organization of the work environment. In other words, technological development, training in the new technology, and the organization of work are regarded as an integral whole.

A central concern of the UTOPIA project involves paying particular attention to two factors concerning the adaptation of the new alternative systems into the work-place. The first of these concerns is to find good methods of performing developmental work which does not result in trade union experts taking over the work centrally. The second concern is to avoid designing a new system which results in a division of labour in the work-place which is dominated by the technical and systems experts. The project methodology also stretches beyond an analysis of the power relationship in the work-place and the forces of technology by seeking answers to fundamental questions such as: What are the limits for machines to solve problems? What are the conditions required for computers to be a tool, an aid, in a fruitful interaction between man and machine? How can the notion of skill be adequately defined and measured? What is the connection between computerization and the change in employees' control of the terminology which is distinctive of their trade? The research methodology is also described by the Swedes as being also concerned with a humanistic and linguistic approach to the central problem of language in the conceptual ability of human beings. The researchers claim that it is much more meaningful to carry out research on the work environment associated with new technological systems by using a theory of knowledge as a starting point rather than relying on scientific data logic systems where workers have to fit in with the conceptual ideas of the system's designers.

The ALC is involved in a number of other projects, such as Computers in Banks, Computers in the Retail Industry, Women's Work, Techniques and Alternatives, and Computers in Office Work. Several of these projects are concerned with the role that the humanities can play in research into the effects of computerization on the world of work. Joseph Wiesenbaum,[19] the Professor of Data Logic at MIT in the United States, has continually emphasized that language is central to computerization. Concept formulations are related directly to language itself, and researchers at Arbetslivscentrum are particularly interested in the relationship between skill depletion as a result of computerization, and the part played by language changes in such a process. The core of a data-processing system is the formalized language, which has a definitive precision in terms of *meaning*. These meanings conflict with the living language which has a wide range of applications in strengthening and broadening interpretations and conveying meanings. The demands of precision and formalization inherent in computer systems often require that complicated processes be simplified, and several Swedish studies illustrated the effect this can have on the work environment.

Conclusions

The Scandinavian countries have made more attempts to deal with social and employment effects of new technology than any other Western industrialized country. Admittedly, the progress these countries have made has been considerably helped by the fact that Denmark, Norway and Sweden have all experienced long periods of industrial peace and stability under centralized industrial relations systems where a social consensus has been in existence between the labour movement and a sympathetic Social Democratic government.

What are the lessons that emerge from the Scandinavian experience? First, it emphasizes the importance of the fact that the direction that the new information technology takes depends on political and economic forces within society. Such forces ultimately determine the contours of power within the work-place where the new technology is applied.

Second, the Scandinavian unions have been instrumental in highlighting the interrelationship between legislation, the negotiation of collective agreements, state-sponsored support for trade union education and research, and the way such an interrelationship is linked to the design of jobs and industrial and economic democracy.

Finally, the developments in Scandinavia illustrate that there are *alternatives* to a 'laissez-faire' approach to the new technology, and there is a definite 'technological awareness' on the part of government, industry, trade unions and the general public in Scandinavian countries. This technological awareness is reflected in the unique character of Scandinavian research on the consequences of information technology. Such research embraces issues outside the world of work such as democracy, individual liberties, cultural values, inequality and personal integrity. The attempts that are being made in Scandinavia to develop a 'Scandinavian model' of the future information society merit much consideration from policy-makers in other Western industrialized countries.

7

New Technology and the Future of Work

This book began by pointing out that the dawn of the 'information era' foreshadows dramatic changes in the world of work. We are probably at one of the major turning-points in history, and much of what is to come in the future will depend on *choices* which are made in the early years of the so-called 'information revolution'. Many of the questions that have been posed throughout this book do not lend themselves to definitive answers; on the contrary, they are questions that will undoubtedly continue to be asked in the years to come as the new information-based technology is gradually introduced into Western industrialized societies: Who will produce what, and how? Which occupations will be radically altered or eliminated altogether? Where are the new employment opportunities to be found in the 'information economy'? What new services will arise to satisfy new aspirations and to take advantage of the new information and communications technologies? Will new technological goods, mass-produced at a low price, enable households and companies to transform themselves into the providers of their own services? What infrastructure is needed to support the new technology? Can we expect jobs to be deskilled, or will there be new skills to be learnt? How will the nature and function of work alter in the years ahead?

Many of the above questions have only been briefly touched upon in this book; any sensible discussion of the choices that are available which will determine the future of work in the information society must be related to several major economic changes that are already taking place.

The Erosion of the Manufacturing Sector

As we saw in chapter 4, traditional manufacturing industry has declined substantially in the Western industrialized societies. Between 1970 and 1981 manufacturing industry in the European countries lost 5.5 million people (of which 2.5 million jobs were lost in the United Kingdom, 1.8 million in the Federal Republic of Germany and 0.7 million in France). This industrial exodus accelerated in the early 1980s, and is mainly affecting the heavy and material processing or manufacturing industries. This suggests that the European countries are gradually moving towards a kind of 'dematerialized' production system where 'light' products and processes will be the foundation of European economies for some time to come.[1]

In the short term at least, one of the major implications of such a transition is the great number and variety of *process* innovations (as opposed to *product* innovations) which are transforming the production system in the Western world, thus contributing to the reduction (saving) of the employed labour forces. According to Christopher Freeman[2] in the present international context of strong competition, process innovations are likely to remain dominant over the next 5-10 years, though energy-based, microelectronics-based and telecommunications-based innovations may also lead to some important new products during this period.

Freeman bases his predictions on the work of the Russian economist Kondratiev, who in 1926 pointed out that economic activity moved in 'long waves' of around 50 years in duration, linked to major pervasive technological innovations. He suggests that the discoveries which will lead to major pervasive innovations may occur decades before their widespread adoption affects the economy. These pervasive technologies, spreading across a wide range of industries and infrastructures, such as steam power, electric power and the internal combustion engine, may produce an economic spurt for several decades, resulting in a massive demand for labour, and involve drastic changes in regional distribution of industry and patterns of investment in it. Whilst governments are able to provide short-term encouragement by helping to build up the appropriate infrastructure and stimulating demand, the long-term potential is only realized when industries themselves have adapted accordingly. How does information technology fit into this perspective?

Freeman claims that information technology is of such a nature that the changes it produces will be the result of process innovation rather than product innovation. It is therefore less likely to generate automatically the extra demand needed to avoid rising unemployment. It is not only in manufacturing where job losses will arise, but also there will be poor prospects for employment growth in utilities, banking, insurance and distribution. Only in personal services are we likely to see much opportunity for employment growth.

Moreover, it is significant that information technology is being adopted during a time of continuing depression by a whole range of large service industries, such as banking, insurance and the civil service. Here, unlike the mechanization of manufacturing production, where increased productivity can generate more profit and further investment, the office sector offers little opportunity for increased profit and further investment, so there is only a shake-out of labour, producing a further slow-down in the economy. As we saw in chapter 4, Gershuny points out that information technology is not being used to develop the necessary *infrastructure* for a whole range of new activities which would provide the necessary stimulation for economic growth. As his research reveals, more and more people are buying capital goods, made by fewer and fewer people, to service themselves at home.

How many more millions of industrial workers in Western societies will lose their jobs in the next few years as the industrial sector continues to decline? Will alternative job opportunities be available in the new 'high-tech.' industries and the service sector of the economy? Employment growth in the services sector is not going to automatically continue in the 1980s, and Gershuny's work shows that employment has already begun to decrease in some traditional branches of services as well as final marketed services. Indeed, he foresees increased *unemployment* in most conventional services and activities, new modes of provision of services (such as do-it-yourself, mutual help, informal economy and voluntary work) and a new generation of services associated with new aspirations and needs (such as more flexible working time, better use of more non-work tied time, better family and social life and greater participation in local and community affairs), new technology (such as information technology and telecommunications) and a changing world economic context (such as transnationalization of economic activities and the emergence of a new world information system). If Gershuny's predictions are correct, four areas are going to be subject to crucial changes: office work, communications, banking and insurance, and education and training.

Changes in Work Time

Changes in the use of work time are already taking place throughout Western industrialized societies. Whilst full-time paid work is still the predominant form of work, part-time and temporary work has expanded considerably since the 1950s as a result of women's entry to the labour market (on average, more than 60 per cent of employed women are working part-time). In addition, of those workers who are in full-time employment, an increasing number are occupied in shift-work. There has also been a shortening of the length of the working day, a growth in annual holidays, and a decrease in the number of years spent in the labour force.

Significant reductions in the length of the working day did not take place until immediately after the first world war, when the length of the standard working week in Britain fell from 54 to 48 hours. This massive reduction was common to all the industrialized countries and the 48-hour week became the norm. The second world war, like the first, witnessed a dramatic change in the use of work time. While there was some reduction in the normal hours of work, the emphasis was on increased living standards through greater social expenditure, and the decline in normal hours of work was more than cancelled by increased overtime for many manual workers.

As far as holidays are concerned, the average number of working weeks in the year in Britain fell by less than 5 per cent between the 1870s and the end of the 1960s; since then there has been a fall of similar magnitude.

As Armstrong[3] points out, it is worth noting that, during the past 35 years, hours reductions and increased holidays in Britain have to some extent been substitutes. During the periods of rapid increase in holidays, weekly hours worked have shown little change.

The number of years spent in the labour force by males has also declined in Britain. As higher education has expanded and the school-leaving age has risen, the average number of years spent by males under 25 years as part of the labour force has declined from 10.4 years in 1891 to 7.3 years in 1981. The participation of workers older than 65 in the labour force has fallen dramatically, from an average of 6.9 years in 1891 to 1.4 years in 1981 – largely as a result of the move towards early retirement. There has been hardly any change over the same period for the 25-64 age group (38.3 years in 1891 and 37.2 years in 1981), and most of the decline in this age group can again be accounted for by the move towards early retirement.

In contrast, the picture of a very large and continuous fall in life hours spent at work for males does not apply to females, at least not as far as paid employment is concerned. The pattern of education followed by full-time work and then retirement applies virtually to all males, whereas in the case of females there are some who have a similar pattern of life/work experience to their male counterparts while others spend very little of their life in paid employment and yet others work mainly part-time. It is therefore very difficult to generalize about female life hours at work, and what data there are should be treated with some caution. The main change in the number of years spent in the labour force by females came after 1931, and more particularly after 1951. From 1931 to 1976 the average number of years spent in the labour force between the ages of 25 and 64 grew from 10 to 23. This is not, of course, the result of changes in the number of years spent in education, or of later retirement, but simply reflects the greater proportion of females taking up paid employment. To a large extent this growth in female employment has consisted of *part-time* labour, usually by married women.

Armstrong suggests that if the statistics on working time for men and women are taken together, and assuming that the results can be extended for a longer period, the implication is that for the post-war period there has been a significant fall in hours worked per employee, both as a result of falling hours for full-time workers and also as a result of a growing proportion of part-time workers. However, the growing share of the population of the main working age engaged in *paid* employment means that the average weekly hours worked per person have shown little change; what has fallen considerably is the time spent in *unpaid* employment within the household.[4]

In summary, the reduction in life hours worked over the past century has been impressive, and there is every likelihood that this reduction will accelerate over the years to come.

Current Trends in Work Patterns

In chapter 4 the evidence suggested that the new information technology is likely to lead to large reductions in the number of jobs. This trend, while there may be disputes about its magnitude, stems directly from one important characteristic of information technology which differentiates it from all previous technologies, namely that it is *both labour saving and capital saving at the same time*. This is true whether we are considering its application in the office environment, the factory, or in any work organization of whatever form. Its diffusion rate throughout various industrial sectors will be uneven but it is only a matter of time before its full force will be felt.

There are very few occupations that will escape the information revolution. The changes that are to be brought about by new technology will have radical effects on work organizations of every form and size. As we saw in chapter 1, the impact of microprocessors coupled with rapid advances in telecommunications will be felt throughout the enterprise, with management at all levels in the organization being equally vulnerable. Enterprises will shrink in size and there will be an increasing trend towards reduced working hours, self-employment, homeworking, and part-time employment. While there is nothing deterministic in the nature of the new technology (in that it offers *choices* relating to how work in the future can be organized), there is a real danger that *if* it is used purely as a means of enhancing managerial control by eliminating jobs and deskilling the work-force, we will be faced with the prospect of a society with a small number of highly skilled technical jobs, large pools of unemployment and those workers who do have jobs will be subject to increasing forms of electronic monitoring and control.

We also noted that certain groups of workers were particularly vulnerable to changes stemming from the new technology, particularly women, the unskilled, older workers and young people seeking to enter the labour market for the first time.

There is little evidence to suggest that the jobs which will disappear from manufacturing industry and offices will be replaced by new jobs which are created in the service sector of the economy, if only because the tertiary and secondary sectors are both automating *at the same time*. Despite the growth in the service sector of the economy coupled with the increase in clerical and office employment in the 1950s-1970s, a large number of occupations in these areas are ripe for automation. Nor is there much support for the view that the new 'information sector' will be a growth employment area. There may well be a growth of jobs in the so-called 'personal services' sector, such as providing people with food, drink, holidays, leisure, sport, heat, light, transport, cleaning and maintenance, but here too there is scope for a considerable degree of automation, and there are limits to demand for such services. Of course, as we saw in chapter 4, an *accurate* forecast of the likely growth/decline of different occupations in different sectors is notoriously difficult to predict, but the prospects for a return to full employment are bleak.

In chapter 1 we were alerted to the strong possibility that we would see an increasing tendency for ('labour segmentation') to emerge, where an increasing number of jobs became less secure, more flexible and increasingly isolated from the external labour market and the wider trade union movement. In sum, it seems safe to conclude that the new jobs that are created will come from information services and particularly from personal services; such jobs will be few and far between and will be nowhere near sufficient to return us to anything resembling the full employment that we experienced during the 1950s and 1960s; and finally, most of the new jobs (with the exception of those requiring very high technical skills) are likely to be inferior in job content and in terms of working conditions.

The trade unions have had limited success in their policy of seeking to influence the direction of the new technology by bringing it within the ambit of collective bargaining through new technology agreements. They have been severely hampered by the fact that microelectronics has emerged during a period of recession, and their bargaining power has consequently been reduced. Most technical change is taking place without any kind of agreement, and where trade unions have succeeded in securing new technology agreements such agreements have fallen far short of their model aspirations. In any case, trade unionists find themselves facing an insoluble dilemma on jobs. If the new technology is introduced then jobs will disappear, but if new forms of automation are ignored then jobs will disappear as the market position of the firms that employ them is eroded by more innovative competitors.

Our debate about the effects of information technology on employment levels could continue *ad infinitum*, and many of the questions that we have asked still remain open. Some of the prospects and possibilities that we have discussed, particularly in the early part of the book, are referring to changes that are on the distant horizon – say in 20 or 30 years time. Nevertheless, even though we are unable to be very *specific* in our predictions of the future of work in the information era, what does the future appear to hold?

Future Work Patterns?

Many of the arguments which have been put forward throughout this book point to a very different form of work organization from the kind we have known in the past. If the present trends are significant, we are likely to see:

1. A situation where full employment cannot be guaranteed,

and where fewer and fewer people are involved in *paid* full-time employment.

2. A manufacturing sector that is smaller in terms of people employed but operating at considerably higher levels of productivity than at present, and more reliance on shift-work and subcontracting.

3. A demand for more highly technically qualified people to service the growing 'telematics' sector as well as more specialists and professionals, but fewer and fewer less qualified workers.

4. Shorter working lives, increasing flexibility in work tasks, more part-time and home-working, short-term contracts based on fees rather than guaranteed life-time employment, and more self-employment.

5. Work organizations in the future will be much smaller both in physical terms and also in the number of people they employ.

6. The boundaries between leisure and work will become increasingly blurred and much more importance will be placed on the 'informal' economy of the home and the community.

7. There will be an increased demand for education at all levels.

8. A smaller earning population and a larger dependent population.

9. Fewer manual jobs and a much smaller (and weakened) trade union movement.

10. More 'self-servicing' in the home and the community.

11. New forms of social organization and government to complement the changes in the organization of work.

The Work Ethic

Perhaps one of the biggest problems facing society in adapting to the changes that are on the horizon is the central importance that *paid* work has occupied in Western societies for the past 150 years or so. The status and identity of individuals in the wider society have been largely influenced by their occupation. Indeed, the terms work, occupation and employment are frequently used interchangeably in everyday speech. It is difficult to conceive of a future where 'work' in the traditional sense of the word becomes less and less important both for the individual and society as a whole. Our self-esteem and the opinion that others hold of us is measured almost entirely by what we *do* rather than what we *are*. Thus an employed person of 60 perceives himself (and is perceived by others) as being of much greater worth than he will be as an unemployed person of 65, 5 years hence. People who have to retire early for medical reasons or who opt to take early retirement as part of a redundancy scheme are immediately seen as having lower status and value in the eyes of others.

Similarly, unemployment and leisure are seen as opposite sides of the same coin but psychologically their impact is totally different: one is feared and the other is eagerly sought. Unemployment has disastrous personal implications for the individual, and is identified with rejection, uselessness, dependence on others, laziness and social isolation. Leisure is sought and enjoyed and is equated with self-sufficiency and the ability to make choices. The work ethic defines work as making leisure meaningful and conversely, leisure is seen as making work meaningful. As Barry Jones asserts:[5]

> The exaggeratedly absolute position asserts that work, however debased, is always good, while non-work, however welcome, is almost always bad, a form of incapacity and humiliation. It is as if work were seen as a raft in a shark-infested sea: being on the raft means safety and security, being in the sea means disaster; the idea of moving on and off the raft voluntarily has no appeal. There are winners and losers: no intermediate position is possible.

The changes which will come about as a result of the introduction of information technology will provide a major challenge to the work ethic. Jobs will become increasingly difficult to find: an increasing number of economists are claiming that full employment will never return to Western societies. The International Labour Organization (ILO) estimates that 1,000 million new jobs would have to be created between now and the year 2000 to achieve full employment worldwide. Moreover, recent studies of unemployed workers carried out by the Science Policy Research Unit at the University of Brighton[6] suggest that even if large numbers of jobs *were* to be created in the future, any policy which sought to redistribute work in the future would have to take into account the fact that the typical skills and work history of the majority of the unemployed are *not* those demanded in the jobs which seem most likely to be created in the near future, nor are they of the kind which could be shared without creating financial hardship. On present trends, job creation seems most probable around new technology-based activities and unmet social needs, and job sharing around highly-paid work: but most unemployed men possess few skills *relevant* to such work, and few possess the skills appropriate to a greater participation in the 'informal' economy.

Somehow or other we will have to find alternative solutions to cope with the radical changes that are on the horizon. There has been no shortage of suggestions, ranging from increased government expenditure, reducing the length of the working week, job sharing, early retirement, workers' co-operatives and reducing the proportion of time that people spend as part of the labour force during their lifetime. Some of these options have already been adopted to a small degree over the past decade; but these have usually been seen as ad hoc 'temporary' responses to rising levels of unemployment rather than as serious long-term measures to adjust society to a prospect of continuing job losses.

If we are to avoid society being torn apart in a bitter struggle between those who have jobs and those who do not, what can be done?

Increased Government Expenditure?

This option is widely canvassed by the Left and many economists who subscribe to a Keynesian perspective. Indeed, nearly all trade union federations in Western Europe believe that increased government expenditure on job creation is an urgent priority. The policies and documents produced by Western European trade unions[7] are distinctly Keynesian products. Running through them are commitments to full employment, growth fuelled by demand management, a necessary state role to assist market performance and aid the social responsibility of companies, etc.

Governments can do much to create employment by increasing capital spending on the services infrastructure of society, for example rebuilding and modernizing roads, railways, sewers, hospitals, homes and schools, and thus provide more jobs in both the private and public sectors. They can also become bigger employers themselves by making more money available to employ more teachers, hospital staff, etc., in labour-intensive industries such as the health services and education. Services to the public can be improved in a large number of sectors where the government has a primary responsibility. In addition, governments can play a major part in, for example, sponsoring research and development to industry, grants for starting new businesses, grants to enable restructuring of industry to take place or educational grants of various kinds.

Governments throughout Western Europe have been doing just this for many years, although many would argue that they ought to be spending much more to assist job creation. Few people would argue that government intervention in the labour market is not desirable, but there are limits to the number of jobs that can be created or maintained. Beyond a certain point, diminishing returns set in and efficacy begins to decline and costs begin to rise.

For example, the Select Committee of the British House of Lords[8] calculated that insulating all the houses in Britain might occupy 3,000 people for 10 years. A large public works programme on railways, roads and sewers, according to the Committee, might need 62,000 people per annum. In order to create one million jobs in Britain over £10,000 million of public expenditure would be required, and even if the British public were prepared to bear an increase in income tax from 30 to 38 per cent in order to finance this, unemployment would still stand at around 3 million by the end of the 1980s because of the extra numbers joining the labour force! All this is (not) to argue that more public expenditure is not desirable, but simply to describe the magnitude of the problem. Of course, there is a great deal that governments can do and ought to be doing in coping with the employment effects of information technology, but the unemployment problem cannot be solved by increased government expenditure on job creation *alone*.

Reducing Working Hours?

Another solution for alleviating the unemployment threat posed by the new technology which is becoming increasingly prominent on the bargaining agenda of the new technology agreements that the trade unions are seeking to negotiate is a reduction in the length of the working week. As the British TUC publication *Employment and Technology* put it:[9]

> There will be a need to consider, at every stage, the opportunities for linking technological change with a reduction in the working week, working year and working lifetime. This should be seen not just in the context of sharing out less work to avoid increasing

> unemployment, but in the more positive light of setting job-creation targets for enterprises and accompanying them by measures which enable existing workers to take some of the benefits in the form of increased leisure whilst providing more jobs.

The 40-hour week remains the norm in most advanced industrialized economies, although unions in Western Europe have pressed for shorter hours since 1979 in the hope of creating new jobs. So far, they have won only a series of small advances rather than sweeping victories. Linked to these demands are other proposals designed to make room for more jobs, including longer holidays, early retirement, sabbaticals, job sharing and short-time working schemes. Following the worst post-war industrial dispute in West Germany by 2.6 million engineering workers organized by IG-Metall, which resulted in an arbitrator's award of a basic 38½-hour working week from April 1985, the General Secretary of the International Metalworkers' Federation, Herman Rebhan, claimed that:

> The 40-hour week is now in the dustbin of industrial history. . . . Over the next few years we shall see working time continually nibbled at and reduced. By the late 1980s most of the industrialized world will be on 35 hours or less, and in a decade we shall see the first 30 hour agreements.

While Herman Rebhan's time-scale might be somewhat optimistic, he does nevertheless have history on his side. Working hours have been falling since the early nineteenth century, and there is no sign of an end to this decline. However, the West German case does illustrate that trade unions have so far enjoyed only limited success in their objective of cutting working hours to save jobs. IG-Metall had hoped that a 35-hour week would create 50,000 jobs. Any employment created by the 1½-hour cut will clearly be smaller, and the effect of the strike may even cost jobs.

The 40-hour week may take some time to die. It lives on in Denmark, Ireland, Austria, Greece, Spain, the USA, some Japanese companies, large parts of Sweden and Finland, Italian private industry, West German sectors that were not involved in the IG-Metall dispute (except the chemical industry), and even some parts of Belgium, the Netherlands and Britain, where it has otherwise been widely breached. In France, the Mitterand government is widely seen as having mis-handled the move to a 39-hour week in 1982 – without loss of pay. The French government claimed that it saved 70,000 jobs, while the employers claimed that it had saved only 30,000 jobs in the short term, and complained that the long-term damage to competitiveness would be severe and would result in a loss of jobs.

As noted in chapter 4, it is by no means clear whether reducing working hours necessarily results in the saving of jobs or jeopardizes them by damaging competitiveness. The conundrum is that if productivity is increased to offset the cut in hours, no jobs are created; if it is not, employers' costs rise – unless they are offset by cuts in real pay – because companies either complete orders more slowly or have to use extra overtime to finish orders on time. One estimate[10] suggested that overtime working in British manufacturing industry represents half a million jobs. The 1982 British government publication *Social Trends* claimed that a curb on excessive overtime (8 hours a week) would create another 100,000 jobs.

The issue of cuts in working hours is linked with a debate on flexibility in the labour market. Faced with demands for shorter hours, employers are insisting in return on the right to deploy workers more flexibly between tasks, and adjust the size of their labour force more quickly and easily. Some commentators argue that short-term artificial creation of jobs will increase costs and either lose orders or, at a later date, lead companies to rationalize, improve productivity and accelerate the introduction of labour-saving new technology!

In any case, there are also divisions on the union side. For instance some Scandinavian unions can see the social benefits of shorter hours, but are not impressed by the job-saving argument. Among the unions which think they can save jobs, there is a split between those like the Dutch unions and the CFDT in France which accept income-sharing, and others which insist that shorter hours must be without loss of pay in order to maintain their reflationary effect.

Many industrial commentators suggest that it is likely that the working week in Britain for manual workers will gradually shorten to 35 hours and that overtime will gradually be reduced; this will come about not as a result of legislation but more likely as a result of collective bargaining under increasing automation. The hours worked in industry will probably continue to be greater than those worked in the services, but the proportion of people in services will keep the average down. In reality the weekly hours of work will be 30 hours, because by the mid-1990s the part-time work-force is likely to be close to 30 per cent of the total labour force (instead of around 20 per cent as at present) and this will bring down the average working week to 30 hours. If employers are unable to make up their labour shortfall by overtime, they might well seek to take on short-term part-time workers rather than increase their full-time work-force. Certainly the number of part-time workers looks likely to increase, but it is doubtful whether there is necessarily any causal connection between an increase in part-time workers and cutting working hours.

But it does seem likely that a reduced working week will lead to greater utilization of capital equipment by more shift-working, and increased flexibility in the organization of work by the increasing use of individualized contracts designed to fit in with the individual worker and the demands of the organization. Both of these developments, by making more efficient use of the capital equipment of an organization, could produce more but smaller jobs because the added productivity would pay for them.

In sum, we can expect working hours to be reduced in the next few years but there is considerable doubt whether this trend in itself will save all that many jobs. Certainly, reduced working hours is not such a panacea for saving existing full-time jobs as some trade unions would claim.

Redistributing Work: Reducing Working Lifetime?

Another suggestion which has been put forward to alleviate the unemployment threat of the new technology is the notion of redistributing work amongst the population by reducing the length of people's working lives. Supporters of this view argue that a much shorter working lifetime should be seen as the norm rather than the 50-year model which is still regarded as the model for working-class males – although not for women or those who receive further education of one kind or another.

As we saw earlier in this chapter, over the twentieth century the duration of working lives has been reduced for many people, but more often than not in the early years – either by education or by enforced leisure as a result of youth unemployment. Could this be accelerated by legislation or otherwise?

Not only would the cost of introducing a universal reduction in the retirement age be expensive, but also there is little evidence that it would be enough to restore the economy to anything approaching full employment. All it would do is to redefine the normal working life and therefore the 'working population' – in the sense of those who are eligible and anxious for a job. As such, although it would have some effect on unemployment, it is by no means a panacea or some grandiose work-sharing plan but an economic compromise. In any case, many would still see such a measure as an unacceptable restriction on their right to choose to continue to work in their later years if they so wished.

Retirement at the age of 55 works well enough in Japan because the Japanese family structure adjusts to it. However, if compulsory early retirement were to be introduced in other Western countries it could cause enormous psychological problems of boredom and frustration – unless people were able to accept recurrent education and were able to break away from the work ethic and value free time for its own sake.

Nevertheless, we can expect to see a continuation of the trend towards a gradual reduction in the average length of people's working lives, largely because of more flexible and more 'portable' pension schemes; the years from 16 to 19 being increasingly seen as a period of education, training and work experience; more older workers being able to work on a part-time basis without loss of pension rights; and the likelihood that more and more people will be able to take periods off work for raising children, undergoing retraining or attending educational classes without the loss of pension rights.

The idea of reducing people's working lives emphasizes that the length of people's working lives could not only be a means of sharing *work* but also a means of sharing *income*. In other words, *paid* work should not be considered as the only means of earning one's living, even if there were enough jobs to go round. This is why it is important to consider *all* forms of work, whether paid or unpaid, because many forms of unpaid work either provide extra money, save money, or substitute for money.[11]

The Idea of a 'Time-Bank'

Andre Gorz[12] maintains that society can only solve the problems of job loss stemming from information technology by moving towards the idea of a 'time bank', or a social account of the number of hours worked which allows everyone to lend or borrow time from society while being guaranteed a lifelong minimum income. According to this notion, the only condition would be that everyone should work a minimum number of hours in their active lifetime – a number which may vary with technological progress or politico-economic choices. He quotes the work of the Echanges et Projets group in France[13] centred around Jacques Delors, which advocated that time should be freed.

This would not be done simply by imposing such a policy by legislative or bureaucratic means but by abolishing compulsory working hours (even at school) entirely, so that each individual has real freedom to choose when he or she wants to work. The Echanges et Projets group stressed the need to escape from 'the universal productivist injunction' or 'the system of prefabricated timetables'.

> Every wage earner must be given the possibility of reducing his or her own work time and pay; the employers should have the right to reject this only in a limited number of specifically defined and controlled circumstances.

Gorz dismisses accuzations from his critics that the idea of a time-bank is no more than a Utopian dream. He cites an example in Germany where a limited form of such an idea is already in operation.

> In the Beck stores in Munich, employing 700 people, everyone is able to choose the monthly amount of work that suits them best. Their choices are reviewed each month at staff meetings, where employees monitor the situation and allocate work time by reconciling the needs of the job with each individual's preference. This is self-management of time – and of work too. There is also a daily meeting with the departmental head to negotiate the required hours of attendance. Everything is possible, providing that someone volunteers to stand in for another person on unscheduled leave.

Whatever the merits of the time-bank idea, it is certainly true that successive opinion polls show that there appears to be a gradual move on the part of wage earners to preferring time over money. In Sweden for instance, in the space of 22 years, the proportion of those reported to attach more importance to their work as opposed to their free time has fallen from 33 per cent to 17 per cent. In 1978, a French opinion poll carried out for *Le Nouvel Observateur* by SOFRES asked a sample of working people representative of the population as a whole including those in retirement or not working the following question: 'In ten years' time, thanks to technological progress, you could either earn twice as much or work half as much. What would be your preference?' Sixty-three per cent responded that they would prefer to work half as much and 37 per cent preferred to earn twice as much (13 per cent did not respond).

Moreover, as we noted earlier, an increasing number of people are working on a part-time basis and it is generally believed that the productivity of two people working part-time (for example in the odd instance where a husband and wife team share the same full-time job) is much higher than that of one individual working full-time. Gorz claims that if the same proportion of part-time workers existed in France as in West Germany (14 per cent instead of 9 per cent in 1978) there would be 500,000 fewer unemployed.[14]

A shorter working week or an increase in part-time work are far from being the only, or the best, means of freeing time as productivity increases as a result of information technology over the next two or three decades. As average working time falls to 30, 25, or even 20 hours per week it will be essential to introduce even more flexible arrangements for working. For example one can envisage 'retirement advances' available at any age in return for an equivalent postponement of final retirement; sabbatical years and perhaps allowing people who have worked more in previous years to stop or reduce work for a year without loss of earnings.

The time-bank idea certainly has its appeal as a means of coping with the worst features of the information technology revolution, but it will involve dramatic changes in the role of the state, the education system, the work ethic, the family, trade unions, the use of resources, as well as notions of equality, income and wealth distribution, and democracy. It is far beyond the scope of this book to discuss the detailed implications of such an idea but there is no doubt that the subject will receive a great deal of attention over the next few years as the implications of increasing automation come to occupy a prominent place on the political agenda of all Western democracies.

New Forms of Job Design

Not only do we have to have a vision of how society can adjust to all the changes stemming from information technology and how work can be redefined and redistributed, but we also need to consider ways of redesigning jobs to make work itself more intrinsically satisfying and using the new technology in such a way that it matches human ability and *fosters* skill, rather than seeking to eliminate it. As we saw in chapter 3, there has been a long tradition of Taylorism in manufacturing industry whereby jobs were progressively fragmented and operator skills were reduced wherever possible.

Howard Rosenbrock, Professor of Control Engineering at the

University of Manchester Institute of Science and Technology, is fond of recounting the case of a plant making electrical light bulbs, which was almost completely automatic:

> Here and there, however, were tasks which the designer had failed to automate, and workers were employed, mostly women and middle-aged. One picked up each glass envelope as it arrived, inspected it for flaws, and replaced it if it was satisfactory: once every 4½ seconds. Another picked out a short length of aluminium wire from a box with tweezers, holding it by one end. Then she inserted it delicately inside a coil which would vaporise it to produce the reflector: repeating this again every 4½ seconds. Because of the noise, the isolation of the workplaces, and the concentration demanded by some of them, conversation was hardly possible.

Rosenbrock's example could be matched by countless other examples of degraded work taken from plants and industries in industrialized countries, but he highlights the way that engineers and designers frequently perceive human capabilities in the work environment by considering how first-year engineering students might react in a design exercise to the challenge of automating the second job described above. After considering all the possible solutions, one student might suggest using a small robot fitted with optical sensing. The lecturer, being an experienced engineer, would probably reject such a proposal because it would involve using a complicated device to meet a simple need, and would offend the engineer's 'instinct of workmanship' – the sense of economy and fitness of purpose. After further discussion, it might emerge that a small number of new plants could be built which would incorporate just such a robot, because the development costs could be outweighed by the resulting efficiency. Yet this solution might still offend the 'instinct of workmanship' of the robotics specialist. 'Why bring in a universal robot that might mean using a machine with many abilities to do a single job that might require only one ability?' The robotics specialist might then turn his or her attention to rearranging the production line to take over some other task which more nearly suited its abilities. As Rosenbrock concludes:

> As engineers we should not rest happy with the design while a gross mismatch existed between the means we were employing and the tasks on which they were employed . . . there is a curious discrepancy here between the apparent attitudes to robots and to *people*.

The American literature abounds with examples which illustrate the thinking that sometimes lies behind Tayloristic ideas of job design.[15] The irony is that is 'respectable' to tackle the misalignment between human abilities and the demand of some jobs by carrying the process of automation to the point where human labour is eliminated.

For example, robots are now being introduced to carry out tasks such as paint-spraying and repetitive spot-welding – all of them tasks which are generally considered to be dirty, dehumanizing, dangerous and repetitive. But why were these jobs which were formerly done by human operators so designed in the first place? It was precisely because they were the result of a long process of fragmentation and simplification to fit in with the imperatives of Taylorism. It is ironic that the early applications of robotics in many cases allow us to automate such tasks out of existence. Whilst most of us would applaud the fact that robots are now taking over such demeaning tasks, it does not alter the fact that technology is still seen by many managers, designers and engineers as a means of making human beings subservient to the dictats of the machine – with the intention of eliminating human judgement and discretion wherever possible. There is no reason why engineers and managers should continue to follow this path of subordinating work to the machine, fragmenting work into rigid components which result in tedious and demeaning tasks for millions of working people until the best thing that can be done with the jobs that remain is to automate them out of existence! A path can be followed through which human skill *can* be preserved – not necessarily by becoming fossilized in old patterns, but by evolving into new skills in relation to new machines and systems. This can only be done, however, at the beginning of the process of technological application. it is at that time when *choices* can be made. If new technology takes the alternative path of being used to draw upon and foster human skills instead of seeking to eliminate them by a gradual process of simplification and fragmentation, then there might not be such a need to consult social scientists about suggestions for some form of job redesign which might alleviate the monotony or the pressure inherent in so many jobs!

The Politics of Design

The *design* of the technology can either limit the organization of work around it or it can be used to provide alternatives for management in the way in which it is used. For example, a research team at the University of Manchester Institute of Science and Technology under the direction of Rosenbrock,[16] is currently working on the development of a flexible manufacturing system with the intention of creating a system which allows for a 'computer-aided' craftsperson who would be responsible for the whole job of making a part, once it had been designed. He or she would make the first part in a batch using the numerically controlled machine tools and interacting with them through computer interfaces. His or her operations would be recorded and repeated automatically to make the remaining parts of the batch. The robot, too, would be programmed by the operator.

The system is reported as having the technical advantages of easier programming of the robot because the full three-dimensional situation is evident to the operators, and also rendering separate verification of programmes unnecessary because making the first component automatically verifies the programme which results.

In addition, the system is more flexible because the operator skill is available to deal with the multitude of difficulties and special situations which arise. Such flexibility is of positive economic benefit in allowing a greater deal of product variety (for example tailored specifically to the requirements of different markets) and quicker production changes (compared with conventional mass markets).

The above example provides evidence that it *is* possible – providing that the appropriate objectives are set and appropriate choices made by management, systems designers and trade unions/employees – to incorporate skilled work and facilitate decision-making on the shop floor in concert with the introduction of new technology.

It is important to recognize that the effect of new technology on the work-place cannot be analysed in isolation from the political context of the design process itself. The *choice* of technology is dependent on social values and political interests. There is no inherent reason why designers of new technology systems should produce machinery and computer systems which deskill and trivialize the tasks of those that work with them. In all probability, many engineers and designers are usually unaware that social choices have already been made in the design process itself, largely because it is often carried out in isolation from the particular point of application, without reference to payment systems, skill hierarchies, established working practices and the politics of the work-place. The Swedish projects which we discussed in the previous chapter, where social scientists and workers who will eventually use the new systems work alongside the designers, have much to commend them in this regard.

As Bjorn-Anderson and Hedberg[17] concluded following a study of automation in banking:

> The scope of the design process must be broadened, users' participation must increase, and the political aspects of designing must be recognized and dealt with. Training, diversity, and consciousness raising, together with more diverse performance measures and supportive reward systems can enable and encourage design teams to design with both human needs and technological possibilities in mind. . . . Users must improve their resources to influence design processes, mere participation is not enough. Increased self-reliance, access to expertise, and legislative support which grants users the right to share design activities can strengthen users' influence over the design processes. . . . Participative designing can contribute to better designs when design teams share the major objectives. When there are many conflicting interests, information systems designing should be seen also as a political process, and designs should emerge as compromises between different organizational sub-groups.

Even in the case of new technological systems which have been designed without any regard to the context of the work-place where they are to be applied, managers can, and in some cases do, mediate in the process of changing technology, both in regard to organizational and technological choices. All too often, the application of new technology revolves round the syndrome of 'the engineers supply the technology, we deal with the consequences', and this can prove to be a major obstacle for managers and workers to overcome.

Finally . . .

We are still in the early stages of the information technology revolution. Despite the rapid pace of development there is still time to make choices which will ultimately determine the kind of society that our children and grandchildren will live in and the kind of work they do. Technology can be used to promote greater economic equity, more freedom of choice, and a quality of life which far exceeds anything we can possibly envisage. Conversely, it can be used to intensify the worst aspects of a competitive society, to widen the gaps between rich and poor, and be used by a small powerful elite to exercise control in an Orwellian fashion over the rest of the population.

We are on the threshold of a technological revolution which is far-reaching and is developing rapidly. It is much easier to allow technological innovation to take its own course than it is to understand or deal with its social consequences. The pervasiveness of this technology, together with the rapid pace of its introduction, highlights the importance of finding ways to deal with its possible consequences for the future of work. So far, few nations – with the notable exception of the Scandinavian countries – have taken positive steps to ensure that technology is used to the benefit of society as a whole instead of being limited only by the laws of the 'free market' and thereby benefiting only the most powerful groups in society. The control and application of new technology is a *political* issue, and the threats and the opportunities it offers, not only in terms of its implications for the future of work, but also for the future of society itself, are too important to be left to market forces alone. Science and technology are not neutral; the direction they take and the uses to which they are put reflects the interests of the dominant groups in society.

This is not the place to embark on a detailed outline of measures that are necessary to ensure that the new technology is used for the benefit of all members of society and not just the privileged few. Nevertheless, a useful start could be made in all Western industrialized countries by adopting an equivalent of the Swedish *Joint Regulations in Working Life Act* to provide that prior information about proposed technological changes should be given by its proponents to appropriate trade unions, and that negotiations should follow to ensure that any proposed economic benefits are shared by the labour force affected. There is also a need for an independent body to monitor technological change and to provide critical and adversarial advice on matters such as job design, work patterns, questions of skill and whether particular forms of technology are appropriate to meet particular needs.

Information takes on a particular significance as computerization and telecommunications systems are gradually introduced. Perhaps one of the biggest dangers posed by the effect of information technology on individual members of society stems from the possibility of bureaucratic power being marshalled at the expense of unorganized and powerless individuals as a result of information gathering and storage; complete personality profiles of individuals can be built up by the use of a conglomeration of data from various sources.

The operation of official databanks and filing systems potentially damages civil liberties in three main ways. First, it does so through the intrusion which may be experienced when information is gathered. Second, civil liberties may be affected when stored information is linked in ways or diverted for purposes not originally intended or authorized. Finally, civil liberties may be threatened when information is transferred out of a databank or filing system to agencies who use it adversely against the subject.

The effective working of democracy depends on the availability of adequate information and the capacity for its independent evaluation. The right to be informed is basic to every person: access to information and the right ot its availability should be vigorously pursued as part of a national information policy.

The challenges posed by the new technology need to be firmly placed on the political agenda of all Western democracies. Quite apart from the various measures that can be taken to tackle the problems of structural and personal adjustment in the new 'information age' (such as enabling people to have greater freedom to move in and out of the labour force, extending access to education, culture and other forms of self-expression, redefining 'work', providing a right to a minimum income for everyone), we need to raise our levels of *consciousness* if we are to take full advantage of all that new technology offers. The most difficult changes are essentially conceptual and relate to people's image of themselves, their place in the world and their goals and capacities.

The problems facing policy-makers should not be underestimated. For example, the elimination of much routine, repetitive, boring and/or physically arduous work, no matter how desirable from an academic, aesthetic, bureaucratic or entrepreneurial perspective, represents a major problem of adjustment for many people whose lives have been built around such employment and who lack the requisite education, cultural conditioning, temperament or personal capacity to make a transition to more stimulating or rewarding activities. The adoption of technology which will abolish dirty, dehumanizing and dangerous work must be welcomed unequivocally, but we must also assert the right to choose appropriate types of technology at our own pace, and to express a preference for forms of technology which *enhance and extend* human capacity, dignity, diversity and understanding.

Many of the issues which have been highlighted throughout this book are not new. The questions that have been asked about the future of work under increasing automation have been posed from one technological wave of innovation to the next at various times during the past 200 years.

Undoubtedly, they are also questions which will be asked again as technological innovation progresses in the years (and decades) to come. This is why some readers may be disappointed that no definitive answers to many of these questions have emerged. Nevertheless, just because many of these questions remain open is not to say that we should let the future of work take care of itself. We are still able to anticipate the possible consequences of new technology and identify the choices that can be made which might ultimately determine its future direction.

Notes

Introduction

1. A 'bit' is a unit of information – the abbreviation of binary digit, one of two values used in binary notation (1,0).

1 The Impact on the Enterprise

1. J. Rada, *The Impact of Microelectronics: A tentative appraisal of information technology*, ILO, Geneva, 1980.
2. B. Lamborghini, in G. Friedrichs and A. Schaff (eds.) *Microelectronics and Society: for better or for worse* (A Report to the Club of Rome), Pergamon, Oxford, 1982.
3. Lamborghini in G. Friedrichs and A. Schaff, op. cit., p. 131 (see note 2).
4. Lamborghini in G. Friedrichs and A. Schaff, op. cit., p. 135 (see note 2).
5. *Financial Times*, 17 November 1983.
6. E. Arnold, Information technology as a technological fix: computer aided design in the United Kingdom, in G. Winch (ed.) *Information Technology in Manufacturing Processes*, Rossendale, London, 1983, p. 37
7. E. Arnold, op. cit., p. 39 (see note 6).
8. A. Francis, The social effects of CAE in Britain, *Electronics and Power*, January 1983.
9. S. Rothwell and D. Davidson, Training for the new technology, in G. Winch (ed.) op. cit., p. 73 (see note 6).
10. P. Dawson and I. McLoughlin, Computer technology and the redefinition of Supervision. Paper presented at the British Sociological Association Conference, University of Bradford, April 1984.
11. Lamborghini in G. Friedrichs and A. Schaff, op. cit., p. 155 (see note 2).
12. See H. Lundgren, *Companies within Companies*, Management Media, Box 5157, 10244 Stockholm.
13. A. Francis, op. cit., p. 73 (see note 8).

14. The *Financial Times* reported in June 1984 that a new device had been invented which would enable small batch manufacturing to be carried out no more expensively than mass production.
15. R. P. Rumelt, The electronic reorganization of industry. Paper presented at the Global Strategic Management in the 1980s Conference, London, October 1981.
16. A. Francis, op. cit., pp. 73–4 (see note 8).
17. J. Rubery and F. Wilkinson, 'Outwork', in F. Wilkinson (ed.) *Dynamics of Labour Market Segmentation*, Heinemann, London, 1981.
18. *New Society*, 1 March 1982.
19. *Financial Times*, 1 September 1983.
20. J. Child, New technology and developments in management organization, *Omega*, vol. 12, no. 3, 1984, p. 218.
21. See the results of a survey which was carried out by the Institute of Administrative Management which was reported in the *Financial Times* on 10 April 1983. The survey showed that, among 180 UK companies, administrative and managerial costs had risen by 4 per cent in real terms during the 5 years to 1981.
22. J. Child, op. cit., p. 219 (see note 20).
23. C. Handy, *The Future of Work*, Basil Blackwell, Oxford and New York, 1984.
24. C. Handy, op. cit., p. 82 (see note 23).
25. See in particular, C. R. Littler and G. Salaman, *Class at Work*, Batsford, London, 1984; C. R. Littler, *The Development of the Labour Process in Capitalist Societies*, Heinemann Educational Books, London, 1982; R. P. Dore, *British Factory–Japanese Factory*, George Allen and Unwin, London, 1973.
26. W. Brown, Britain's unions: new pressures and shifting loyalties, *Personnel Management*, October 1983, p. 49.
27. J. Lloyd, Wages: the battle for a flexible future, *Financial Times*, 4 April 1984.
28. T. Nuki, The effect of microelectronics on the Japanese style of management, *Labour and Society*, vol. 8, no. 4, October–December 1983.

2 The Automated Office

1. A. J. Krowe, Enterprise systems: updating the office of the future, in *Best's Review Life/Health Edition*, A. M. Best Co., Oldwick, New Jersey, November 1980, pp. 32–38.
2. Perhaps the most useful book for explaining the new office technology in layperson's terms is M. Peltu, *The Electronic Office*, Ariel Books, London, 1984. This book was published to complement the BBC television series of the same name which was broadcast in the early part of 1984. Malcolm Peltu is a consultant to the EEC on human and organizational aspects of office systems.

3. M. Peltu, op. cit., pp. 120–121 (see note 2).
4. I. Barron and R. Curnow, *The Future with Microelectronics*, Frances Pinter, London, 1979, p. 151.
5. J. M. McLean and H. J. Rush, *The Impact of Microelectronics on the U.K.: a suggested classification and illustrative case studies*, Science Policy Research Unit, University of Sussex, 1978.
6. See D. Werneke, *Microelectronics and Office Jobs: the impact of the chip on women's employment*, ILO, Geneva, 1983.
7. *Datamation*, Los Angeles Technical Publishing Corp., November 1979.
8. A. Pollack, When computers don't work, *New York Times*, 23 October 1981.
9. H. Braverman, *Labor and Monopoly Capital: the degradation of labor in the 20th century*. Monthly Review Press, New York, 1974.
10. By far the best account of the development of Taylorism in the advanced economies can be found in C. R. Littler, *The Development of the Labour Process in Capitalist Societies*, Heinemann Educational Books, London, 1982; see also S. Wood (ed.) *The Degradation of Work?*, Hutchinson, London, 1982.
11. M. Traesborg and N. Bjorn-Anderson, *Microelectronics and Work Qualifications: Report to CEDEFOP*, EEC Centre for Vocational Training, Copenhagen, 1981.
12. J. Evans, in *Microelectronics and Society: for better or for worse* (A Report to the Club of Rome), Pergamon, Oxford, 1982, p. 167.
13. *Office Salaries Directory, 1979–80*. Administrative Management Society, Willow Grove, Pa., 1980.
14. D. Werneke, *Microelectronics and Office Jobs: the impact of the chip on women's employment*, ILO, Geneva, 1983.
15. *Business Week*, 3 August 1981.
16. E. N. Glenn and R. L. Feldberg, Degraded and deskilled: the proletarianization of clerical work, *Social Problems*, Society for the Study of Social Problems, Buffalo, October 1977, pp. 52–64.
17. E. Bird, *Information Technology in the Office: the impact on women's jobs*, Equal Opportunities Commission, Manchester, September 1980.
18. N. Bjorn-Anderson, *Are 'Human Factors' Human?*, Paper presented to the Conference on Man – Machine Integration, London, 1984.
19. H. Downing, Word processors and the oppression of women, in T. Forester (ed.) *The Microelectronics Revolution*, (ed.) Basil Blackwell, p. 275, 1980.
20. For another example which illustrates this loss of social contact in office work, see the case study on 'Word processing in a marine engineering consultancy: Y-ARD', in D. A. Buchanan and D. Boddy, *Organizations in the Computer Age*, Gower, Aldershot, 1983.
21. *Entreprise et Personnel, L'évolution des métiers de secrétariat consécutive a l'introduction de la bureautique dans les entreprises*. Report prepared by H. Douard and C. Gillot for le Sécretaire d'État aupres du Ministre du Travail, Paris, *Entreprise et Personnel*, 1981.

22. H. R. Bowen and G. L. Magnum (eds), *Automation and Economic Progress: a summary of the report of the National Commission on Technology*, Prentice-Hall, New Jersey, 1966.
23. Occupational employment growth through 1990, *Monthly Labor Review*, U.S. Department of Labor, Washington DC, August 1981, p. 46.
24. M. Porat, *The Information Economy: definition and measurement*, US Department of Commerce, Washington DC, Office of Telecommunications, 1977.
25. E. B. Cox, Prospects for automated tellers, 1981–1990, in *American Banker*, June 1981.
26. R. Matteis, The new bank office on customer service, *Harvard Business Review*, March/April 1979.
27. International Federation of Commercial, Clerical, Professional and Technical Employees (FIET) : *Bank Workers and New Technology*, Euro FIET, Geneva, August 1980; *Insurance and Social Insurance Workers and the New Technology* (Euro FIET, Geneva, November 1980; *Office Technology in Industry* (Euro FIET, Geneva, March 1981).
28. *Electronic Data Processing in the Social Insurance Offices: programme of action*, Forsakringsanstalldas Forbund, Stockholm, February 1981, p. 28.
29. Equal Pay and Opportunities Campaign (EPOC), *Women and Word Processors*, London, 1980.
30. Quoted in D. Werneke, *Microelectronics and Office Jobs: the impact of the chip on women's employment*, ILO, Geneva, 1983, p. 81.
31. E. Bird, op. cit. (see note 17).
32. Quoted in *Negotiating Technological Change*, European Trade Union Institute (ETUI), Brussels, 1982, p. 17.
33. *World Labour Report*, ILO, Geneva, 1984.
34. E. Bird, op. cit., p. 16 (see note 17).
35. D. Werneke, op. cit., p. 97 (see note 30).

3 On the Shop Floor

1. C. R. Littler, A history of new technology, in G. Winch (ed.), *Information Technology in Manufacturing Processes*, Rossendale, London, 1983, p. 135.
2. The most detailed and useful account of Taylorism can be found in C. R. Littler, *The Development of the Labour Process in Capitalist Societies*, Heinemann Educational Books, London, 1982: see also C. R. Littler and G. Salaman, *Class at Work*, Batsford, London, 1984.
3. F. B. Copley, Frederick B. Taylor: revolutionist, *The Outlook*, 111, September 1915.
4. L. E. Davis and J. C. Taylor, *The Design of Jobs*, Penguin, Harmondsworth, 1972.

5. *Work in America, Report of a Special Task Force to the Secretary of Health, Education and Welfare,* MIT Press, Cambridge, Mass.
6. N. A. B. Wilson, *On the Quality of Working life*, Manpower Papers No. 7, HMSO, London, 1973.
7. See for example, R. E. Walton, Innovative restructuring of work, in J. Rosow (ed.), *The Worker and the Job: coping with change*, Prentice-Hall, New Jersey, 1972.
8. J. E. Kelly, *Scientific Management, Job Re-design and Work Performance*, Academic Press, New York, 1982.
9. C. F. Sabel, *Work and Politics*, Cambridge University Press, Cambridge, 1982.
10. D. Jenkins, *The West German Humanization of Work Programme: a preliminary assessment*, Work Research Unit Occasional Paper No. 8, 1978.
11. M. W. Karmin and J. L. Sheler, Jobs: a million that will never come back, *U.S. News and World Report*, 13 September 1982, pp. 53–56.
12. M. S. Katzman, Human resources implications of robots in the United States, *Work and People*, vol. 9, no. 1, 1983, p. 22.
13. M. S. Katzman, op. cit., p. 22 (see note 12).
14. B. Wilkinson, *The Shopfloor Politics of New Technology*, Heinemann Educational Books, London, 1983, p. 4.
15. Science Research Council, Working Party Report, NC Machine Tool Systems, in *Colloquium on Robotics,* Paper No. 8, Glasgow, 30–31 March 1978.
16. See P. Marsh, Towards the unmanned factory, *New Scientist*, 31 July 1980.
17. In Sweden a completely automated machine shop will be in operation by 1987, where the customer will tell a central computer what he or she requires, the computer will design the product and control the machining process for even small batches of components and then inform the customer when the order is ready. *Financial Times*, 14 July 1982.
18. B. Wilkinson, op. cit., pp. 48-54 (see note 14).
19. Cited by H. Shaiken, Automation in industry: bleaching the blue collar, *IEEE Spectrum*, June 1984, p. 77.
20. *The Impact of Programmable Automation on the Work Environment,* Office of Technology Assessment, Washington, DC, 1983.
21. B. Jones, Destruction or redistribution of engineering skills?: the case of numerical control, in S. Wood (ed.), *The Degradation of Work?*, Hutchinson, London, 1982, p. 179.
22. M. Rose and B. Jones, Managerial strategy and trade union response in plant level re-organization of work. Paper presented to a Conference on the 'Organization and Control of the Labour Process', University of Aston, Birmingham, 23–25 March 1983.
23. J. Bessant, Management and manufacturing innovation: the case of information technology, in G. Winch (ed.), op. cit., p. 14 (see note 1).
24. D. Buchanan, Technological imperatives and strategic choice, in G. Winch (ed.), op. cit., p. 72. (see note 1).

4 The Unemployment Threat

1. W. Dostal, *Bildung und Beschaftigung im Technischen Wandel*, Beitrage zur Arbeitsmarkt und Berufsforchung No. 65, Institut fur Arbeitsmarkt und Berufsforchung der Bundesanstalt fur Arbeit, Nuremberg, 1982.
2. C. Jenkins and B. Sherman, *The Collapse of Work*, Eyre Methuen, London, 1979.
3. Quoted in A. Gorz, *Farewell to the Working Class*, Pluto Press, London, 1982.
4. I. Barron and R. Curnow, *The Future with Microelectronics*, Frances Pinter, London, 1979.
5. Equal Opportunities Commission, *Information Technology in the Office: the impact on women's jobs*, 1980; see also Working Women, National Association of Office Workers, *Race Against Time: Automation in the Office*, Cleveland, USA, 1980.
6. ETUI, *The Impact of Microelectronics on Employment in Western Europe in the 1980s*, Brussels, 1979.
7. J. Gershuny and I. Miles, *The New Service Economy: the transformation of employment in industrial societies*, Frances Pinter, London, 1983.
8. J. Gershuny, *After Industrial Society?: the emerging self-service economy*, Macmillan, London, 1978.
9. J. Gershuny and I. Miles, *The New Service Economy: the transformation of employment in industrial societies*, Frances Pinter, London, 1983.
10. V. Walsh, J. Moulton-Abbot and P. Senker, *New Technology, The Post Office, and the Union of Post Office Workers*, London, Union of Communication Workers, London, 1980.
11. Cited in the European Trade Union Institute study, *The Impact of Microelectronics on Employment in Western Europe in the 1980s*, ETUI, Brussels, 1980.
12. C. Offe, Some contradictions of the modern welfare state, *Critical Social Policy*, vol. 3, no. 2, 1982, pp. 7–16.
13. A. Burns, *The Microchip: appropriate or inappropriate technology?*, Ellis Horwood, Chichester, 1981, p. 122.
14. In some countries the health sector is not part of the public sector.
15. J. Gershuny and I. Miles, op. cit. p. 191 (see note 9).
16. J. Gershuny and I. Miles, op. cit. (see note 9).
17. J. Gershuny and I. Miles, op. cit. (see note 9).

5 The Trade Union Response

1. TUC, *Automation and Technical Change*, 1970, p. 10.
2. M. Roberts, The impact of technology on union organizing and collective bargaining, in D. Kennedy, C. Craypo and M. Lehman (eds), *Labor and Technology*, Dept. of Labor Studies, Pennsylvania State University, 1981.

3. *Employment and Technology*, TUC Interim Report, April 1979.
4. *Employment and Technology*. Report by the TUC General Council to the 1979 Congress, September 1979.
5. I. Benson and J. Lloyd, *New Technology and Industrial Change*, Kogan Page, London, 1983, p. 171.
6. *Employment and Technology*, p. 9 (see note 4).
7. *Employment and Technology*, p. 1, para 1 (see note 4).
8. M. Cooley, *Architect or Bee?*, Langley Technical Services, 1980.
9. T. Manwaring, The trade union response to new technology, *Industrial Relations Journal*, 1981.
10. T. Manwaring, op. cit. (see note 9).
11. R. Williams and R. Moseley, Technology agreements: consensus, control and technical change in the workplace. Paper presented to the EEC/FAST Conference on *The Transition to an Information Society*, Selsdon Park, 1982.
12. Reported in *Changement Social et Technologie en Europe*, Bulletin D'Information No. 8, Pool Européen d'Études, Brussels, July–August 1982.
13. T. Manwaring, op. cit. (see note 9).
14. H. Markmann, The role of industrial relations in coping with the labour implications of technical change: a trade union perspective, OECD Conference Paper, 6–8 February 1984.
15. I. Benson and J. Lloyd, op. cit., p 180 (see note 5)
16. Reported in *Negotiating Technological Change*, European Trade Union Institute, Brussels, 1982.
17. ETUI, *Negotiating Technological Change*, p. 62.
18. *European Industrial Relations Review*, no. 92, p. 24.
19. US Department of Labor, Bureau of Labor Statistics, *Major Collective Bargaining Agreements: interplant transfer and relocation allowances*, Bulletin 1425-20, July 1981.

6 Alternative Approaches to New Technology

1. In Sweden, the Social Democrats have headed the government (with the exception of the 6 years from 1976 to 1982) continuously since 1932; Denmark had Social Democratic Prime Ministers from the 1930s until 1982 but they usually based their rule on coalitions; Norway has had a Social Democratic government for much of the post-war period and the SPD in Germany was in office throughout the 1970s up until 1983.
2. See A. Sandberg, *Computers Dividing Men and Work*, Arbetslivscentrum, Stockholm, March 1979.
3. See K. Nygaard and J. Fjalestad, Group interests and participation in information system development, in *Microelectronics, Productivity and Employment*, OECD,

Paris, 1981; and B. Gustavsen and G. Hunnius, *New Patterns of Work Reform: the case of Norway*, Universitetetsforlaget, Oslo, 1981.

4. *Negotiating Technical Change*, European Trade Union Institute, Brussels, 1982.
5. The wording of the agreements was kindly supplied by Bjorn Gustavsen of the Work Research Institutes, Oslo and the Swedish Centre of Working Life (Arbetslivscentrum), Stockholm.
6. C. Gill, Swedish wage-earner funds: the road to economic democracy? *Journal of General Management*, vol. 9, no. 3, pp. 37–60, Spring 1984.
7. See R. Lindholm, *Job Reform in Sweden*, SAF, Stockholm, 1974; S. Aguren and J. Edgren, *New Factories*, SAF, Stockholm, 1979; and S. Aguren, R. Hansson and K. G. Karlsson, *The Volvo Kalmar Plant*, The Rationalization Council SAF-LO, Stockholm, 1976.
8. K. Nygaard, The 'Iron and Metal Project', in A. Sandberg (ed.), *Computers Dividing Man and Work*, Arbetslivscentrum, Stockholm, p. 94, 1979.
9. J. Fjalestad, *Some Factors Affecting Participation in Systems Development*, GMD, Bonn, 1981. Cited in Benson and Lloyd, *New Technology and Industrial Change*, Kogan Page, London, 1983, p. 162.
10. ETUI, *Negotiating Technological Change*, European Trade Union Institute, Brussels, 1982, p. 70.
11. C. Gill, Industrial relations in Denmark: problems and perspectives, *Industrial Relations Journal*, vol. 15, no. 1, pp. 46–58, Spring 1984.
12. In 1983, the Swedish Parliament also passed legislation providing for 'economic democracy' through the introduction of a wage-earner fund system. The trade unions saw the issues of industrial and economic democracy to be inexorably linked together. For an evaluation of this legislation, see C. Gill, Swedish wage-earner funds: the road to economic democracy?, *Journal of General Management*, vol. 9, no. 3, Spring 1984.
13. ETUI, op. cit., p. 76 (see note 10).
14. A co-determination agreement covering central government employees came into force in July 1978 and one covering the banking sector came into force in January 1979. Agreements covering other parts of the public sector, co-operative companies and insurance companies were concluded between 1979 and 1981.
15. D. Chamot, Technology: how European unions cope, *AFL-CIO Federationist*, November 1981.
16. Sweden is the third largest producer of industrial robots behind Japan and the USA, and over 60 per cent of its production is exported.
17. English translation from *Regeringens proposition, Samordnad data-politik*, Prop. 1981/82 : 123.
18. European Pool of Studies, *Social Change and Technology in Europe: Current Events in Scandinavia*, Brussels, September 1982.
19. *Computer Power and Human Reason*, W. H. Freeman, San Francisco, 1976.

7 New Technology and the Future of Work

1. R. Petrella. Paper presented to the British Association for the Advancement of Science, Brighton, 1983.
2. C. Freeman, Keynes or Kondratiev? How can we get back to full employment?, in P. Marstrand (ed.), *New Technology and the Future of Work and Skills*, Frances Pinter, London, 1983.
3. P. J. Armstrong, Work, rest or play? Changes in time spent at work. Paper presented to the British Association for the Advancement of Science, Brighton, 1983.
4. J. I. Gershuny and G. S. Thomas, *Changing Patterns of Time Use,* Science Policy Research Unit, University of Sussex, 1980.
5. B. Jones, *Sleepers Wake! Technology and the Future of Work*, Wheatsheaf Books, Brighton, 1982.
6. I. Miles, Adaptation to unemployment? Occasional Paper No. 20, 1983: M. Jahoda and H. Rush, Work, Employment and Unemployment, Occasional Paper No. 12, SPRU, University of Sussex, 1979.
7. See for example the European Trade Union Institute, *Negotiating Technical Change,* Brussels, 1982; Trades Union Congress (TUC), *Employment and Technology,* 1979.
8. *Report of the House of Lords Select Committee on Unemployment,* HMSO, London, 1981.
9. TUC, *Employment and Technology*, 1979, p. 38.
10. G. Merritt, *World out of Work*, Collins, London, 1982, p.159.
11. C. Handy, for example, in his book *The Future of Work*, Basil Blackwell, Oxford, 1984, has outlined a wealth of ideas for reorganizing work in society, including more part-time work and shorter careers, for more voluntary and co-operative work, for more respect for work in the family and the community, self-employment, etc. However, whilst he correctly emphasizes both the magnitude of the jobs problem and the urgency in finding solutions to it, the scenarios he describes appear to be overly optimistic and rooted in the questionable belief that individuals themselves are able to arrive at their own solutions without government support.
12. A. Gorz, *Farewell to the Working Class,* Pluto Press, London, 1982.
13. L.Cosse (ed.), *La Revolution du Temps Choisi,* Paris, 1980.
14. According to the Eurostat Labour Force Sample Survey, the proportions of part-time workers in France and West Germany in 1981 had risen to 10.26 per cent and 17.16 per cent respectively. Eurostat, Luxembourg, July 1983.
15. Rosenbrock cites several examples in a Work Research Unit Paper in 1981, *Engineers and the work that people do,* WRU Occasional Paper 21, October 1981, e.g. 'The US Rubber Company has even pushed experimentation so far as to employ young girls deficient in intelligence who, in the framework of Scientific

Management applied to this business, have given excellent results,' and 'Mike Bayless, 28 years old with a maximum intelligence level of a 12-year-old, has become the company's NC machining centre operator because his limitations afford him the level of patience and persistence to carefully watch his machine and the work that it produces' (quoted by G. Friedmann, *Industrial Society*, Free Press of Glencoe, New York, p. 216, and Frederick Hertzberg, *Work and the Nature of Man*, p. 39, World Publishing Company, respectively).

16. For more details see H. Rosenbrock, Technology, control, information and work organisation: what are the options? *Work and People* vol. 9, no. 2, 1983, pp 9-13; also H. Rosenbrock, Robots and people, *Measurement and Control*, vol. 15, 1982, pp. 105–112.
17. N. Bjørn-Anderson and B. L. T. Hedberg, *Designing Information Systems in an Organisational Perspective*, North Holland/TIMS Studies in the Management Sciences, vol. 5, 1977, pp. 125–142.

Index